ARCHITECTURAL AND ENGINEERING SALESMANSHIP

ARCHITECTURAL AND ENGINEERING SALESMANSHIP

DAVID G. COOPER, A.I.A.

A Wiley-Interscience Publication
JOHN WILEY & SONS
New York • Chichester • Brisbane • Toronto

Library of Congress Cataloging in Publication Data

Cooper, David G
Architectural and engineering salesmanship.

"A Wiley-Interscience publication."
Includes index.
1. Architectural services marketing. 2. Engineering services marketing. I. Title.

NA1996.C56 658'09'72 78-15367
ISBN 0-471-03642-0

Printed in the United States of America

10 9 8 7 6 5 4 3 2 1

To MY THREE SONS
KEVIN, JONATHAN, and STEPHEN

FOREWORD

My first association with David Cooper was at a conference held for people anticipating the need for new construction. I was so impressed with him and his firm that I had them interview, along with five other firms, for a new 60,000 square foot building. His firm was awarded the contract for the architectural work by a unanimous vote of the board of directors. This was the beginning of a fond friendship and my great admiration for David Cooper.

At the time we employed his firm I felt we were really hiring the firm first, and David second. But during and after construction I realized that it was David we had really employed. After David moved away from the area, there was a marked decline in the success of his former firm in relation to their competition in the area.

I feel that what David is truly saying in this book is that people hire other people—not firms—and that the firm is nothing but the people within it.

I happen to be an educational administrator, but if I were in college studying to become either an architect or an engineer this is the kind of enthusiastic, step-by-step detailed book I would hope to have assigned to me or be fortunate enough to have someone recommend to me. Selling architecture is not something that comes easily to architects and engineers, but we building owners or potential clients often need to be "sold." David's pitch for his profession is refreshing, optimistic, realistic, and grounded in firsthand experience in the marketplace.

MARK F. KEEGAN
Superintendent of School
Dexter, Maine

PREFACE

The practice of architecture and engineering today is changing at a faster rate than we could have foreseen a few years ago. Because of rising costs, owners are demanding new contractual methods, and projects are becoming harder to find. When you do secure a real project, it becomes increasingly difficult to finish the project successfully and keep everyone involved happy in the process. By successful, I mean a project that is finished on schedule, falls within the budget, satisfies all the owners' program requirements, and yields a good profit to the architects and engineers.

If you feel that today it is almost impossible to achieve all these, then your attitude will certainly be reflected negatively in your work and in your clients' opinion of you, which will in turn affect your ability to get continued work.

But if we examine what is happening in the architectural and engineering professions today, we find that, even though the projects are fewer, with more difficult terms to meet, some firms are busier than others and have conquered more than their fair share of the available work in their given area of practice and expertise. The question we must all keep asking ourselves is why this is happening when the national professional trend is one of declining work loads and lower profits and when there is a history of firms going out of business and moving away from membership in the national professional organizations.

As I worked to expand the practice of my firm's New England office, I kept asking myself these questions because I wanted mine to be the firm that was expanding and growing and whose profits were going up. In fact, I wanted it to be the leading architectural firm in the area.

Constant analysis and study of the state of the art led me to observe that busy and successful firms were proficient in three areas of expertise and that I must develop proficiency in these same areas to assure that mine would also be a busy and successful firm, even though the national trend was down. This book is based on the recording of my thoughts and observations as I undertook this study.

Design is the first necessary area of expertise; technical skills are second, and nontechnical skills third. This book therefore relates all three areas to the process of obtaining and keeping clients.

1. *Design* is the discipline in which we receive the majority of our formal training in school, and many firms are strictly design oriented to the extent that they concentrate on design and downplay the other disciplines necessary to create a successful building.

Throughout this book I will try to show the relationship of design to the other disciplines and develop a design philosophy valuable to building owners. For while designers may feel that aesthetic considerations outweigh all others, the owner regards the understanding and fulfillment of his or her program requirements as the essential elements of the design process.

To go one step further, it is difficult to design a cafeteria or food service kitchen if you have never worked in one. It is hard to design a truly efficient teaching space unless you have really studied and practiced the art of teaching. As architects, we must be more receptive to the owner's position and experience. To quote a typical owner who told me that the architect who designed his building received acclaim when the building was published in all the design magazines, "But I wish that he would come and live in the building for a year and see what it is that he really has designed."

2. *Technical skills* to which professional architects and engineers are exposed in college are then developed to greater or lesser degrees as they work in their respective offices. This book will touch on the technical skills with which successful professionals must be fully cognizant to insure the satisfaction of clients. But the real meat of this book will deal with technical skill in relation to the searching out, obtaining, and keeping of new and existing clients, which will be presented as a step-by-step process, each step of which you must master if you are to help your firm grow in a positive and successful way.

3. *Nontechnical skills* are the ones to which few of us ever give much thought. They involve the things we do automatically. They deal with how we relate to others; what others think of us as people, not as technicians; and the vibrations and energy that we generate. A study of this unfamiliar but essential element of professional competence forms a basic part of this book. It is something that cannot be taught in school, but must be learned and developed in the world of work to assure success.

You can be the most competent designer and master in professional skills, but if people do not enjoy your company and want to build a working relationship with you, you will find it difficult to get work and almost impossible to maintain a record of repeat clients. To help you understand this area, I have given examples throughout the book of ideas and techniques that I used in developing my own personal, nontechnical skills—the philosophy of human relations. I hope that by setting down a careful analysis of what has worked for me and for others whom I have had the opportunity to observe closely over a period of time, broad principles will emerge that you can apply to your own professional career.

These techniques were developed when I was working for a very large architectural and engineering firm, and were instrumental in helping the firm's New England operation to grow and expand while its operations in other parts of the country were only holding their own. I have recently become a member of a small general construction company in Florida and find that the philosophy and techniques apply equally to both situations.

The step-by-step procedures set forth in this book—from organizing your personal time in the office, to preparing for the interview, to finally closing the deal and performing successfully for the client—have their roots in one preeminent principle. Success will ultimately depend on satisfying the owner. This book is dedicated to helping you achieve the kind of owner satisfaction that will lead to a healthy flow of new and repeat clients.

The gathering of material and the development of the techniques and philosophies contained in this book took place from 1969 to 1978. During this period of time many people helped me, some knowingly and others unknowingly. It is thus only fitting that credit be given to them at this time.

The first group were my friends and fellow workers with whom I struggled to achieve my personal growth and the growth of my firm. This group includes Erney Hirschoff, Barbara Mair, Edwin Bruce, and Mary Martin.

The second group of people helped take my ideas and produce the finished book. S. Peter Volpe encouraged me to start the book. Dudley Hunt never lost patience with me as I missed deadline after deadline that he had set for the publication of the book. The two most important contributions, however, were made by Barbara Schmader and Susan Braybrook who took my many pages of rough manuscript and converted them to the finished product without losing the feeling or the intent.

The "thank you's" would not be complete without mentioning my many friends in the County and elsewhere in New England who believed in me and helped me with my many projects and ideas.

To all of you my deepest thanks.

David Cooper, A.I.A.

Naples, Florida
October 1978

CONTENTS

ARCHITECTURAL AND ENGINEERING SALESMANSHIP

CHAPTER ONE

WHAT DO TODAY'S OWNERS WANT?

The key to your success rests with the owners and their need to be satisfied with their project. If you can analyze and relate positively and realistically to the needs of the owners, you can then work to satisfy those needs. If you can achieve a history of satisfied owners, this will be the single most important factor in the ability to achieve a healthy flow of work. Assuming that the general level of owner satisfaction with the architectural profession will continue to be low, if you do an outstanding job in service to the client and simultaneously establish and maintain an active, aggressive prospecting program, you will be out in front of most of the competition.

STATE OF THE PROFESSION

Typical Owners' Comments

When we take a look at the typical owners of today, we find that they, like architects or engineers, are affected by the doom-and-gloom syndrome. They believe the day of forced expansion is over and that their current building program may well be their last. They feel we are entering an era of replacement and renovation. But whether new construction is part of the building programs of the future, or the programs will be for replacement or renovation is not important to the discussion; what matters is that owners are still building and will continue to build.

In general, the owners I have talked to are very disappointed with their last building project. Many call their projects disasters. They claim that the buildings came in over the budget and were not finished on schedule; and that the mechanical system operates imperfectly. If we examine these typical complaints more closely, we find that owners often brand their last building ventures as failures for one or several of the following reasons. None is new to people in the professions, but to improve the client-profession relationship and give the owners a better understanding of the building process, it is worth examining the following lists and then taking a look at what lays behind the specific problems.

1. ARCHITECTS OR ENGINEERS FAILED TO PERFORM

   Cost overrun
   Building completed late
   Building does not function as programmed
   High maintenance cost for operation

2. CONTRACTOR DID NOT PERFORM

   High costs
   Reduced quality
   Late delivery
   Punch list not completed

3. MANUFACTURERS LET EVERYONE DOWN

   High costs
   Late delivery
   Sold products not in production
   Salesperson made promises the company did not keep

We have all heard these complaints repeatedly and tend to shrug them off saying we can do little about them. But the time has come when we must start doing something about them. The construction of his or her building may well be the most important achievement of the owner's career, and the complaints we try to dismiss are to an owner major disasters. To people in the industry who handle many projects throughout a year, these delays, cost overrun, and lack of performance are often accepted as the routine occupational hazards of the construction business. But if we do not try to correct them, we shall find that the owners will do so themselves, and in so doing, completely change the roles of the members of the construction team.

Some years ago, I conducted an in-depth study of the needs and desires of owners in relation to the construction process for the W. K. Kellogg Foundation. At that time, I was serving in a part-time capacity as the organizer and director of the Kellogg Foundation-funded New England Construction Education Program. Operating at the University of Vermont, this program's objective was to define the needs of the construction industry in New England and then develop educational programs to meet them, which would be held in key locations throughout the six states. At the heart of the organization was a fifty-member, horizontally organized advisory council. That is to say the council included representatives from all the professions and disciplines involved in the construction industry. The list included owners, government agencies, large contractors, small contractors, subcontractors, bankers, lawyers, union officials, material suppliers, architects, engineers, construction workers, and educators from the state universities. The council, which met for a day four times a year, worked to define the needs of the region and then to assign priorities in trying to meet them. From this list, the project staff would develop programs and take them back to the council for their review. Once the council approved the programs, they were finalized and put on the schedule for presentation throughout the six New England states. This interdisciplinary council proved very effective in defining the construction education needs of the New England states.

Our first program was called "project leadership." It dealt with the problems of getting the people in the field to work together successfully to get the building built. This was presented eight times throughout the region as a day-long program to an audience of architects, engineers, contractors, and owners.

After supper, the participants would reassemble for a panel discussion on problems at the job site. An architect, contractor, owner, clerk of the works, and a contractor's superintendent would always be on the panel. When we organized these panels, the participants were specifically invited to talk about their own experiences in trying to get buildings constructed. We hoped in this way to define some of the problems more clearly so that we could attack them effectively in future seminars. We expected the discussion to be lively, with the architect and owner taking one side and attacking the contractor for causing all the problems. With this as the background, we started the first seminar.

As we got into the discussion, it quickly became evident that the owner-architect versus contractor conflict was not going to materialize, but rather that the lines of conflict were drawn between the owner-contractor versus the architect. To my astonishment, the architect was always the "bad guy," and this confrontation dominated all eight seminars. Now this is not to say that in all cases the criticisms of the architect were justified. In some cases, the architect was certainly used as a scapegoat. The point is not so much whether the criticisms were valid, but that the architect was always cast in the role of incompetent or "bad guy," who spent his or her time trying to evade the consequences of his or her own incompetence. This of course is contrary to everything the architect is taught to believe about himself or herself in school and in practice. But in the real world, the architect is clearly being identified as the person causing the problems. The contractors in these seminars defended their position by saying that they always built what the architects' drawings showed, while the owners maintained that they gave the architects their programs and received in response buildings that were beset with problems.

Specific Problems with Architects and Engineers

Some of these problems were mentioned during the eight seminars:

1. DO NOT KNOW COSTS. Projects are constantly over the budget, but when contractors become involved, they can usually show a cheaper way.

2. DO NOT KNOW DETAILING. Detailed work is often expensive to build, hard to maintain, or fails to function as designed. Owners often wonder if architects really know how a building and materials go together.

3. HAVE A SUPERIOR ATTITUDE. They say, "Give me your money and your program and come back when the building is finished. In the meantime, don't bother me." But the finished building is often not what the owners want.

4. MOST OF THE PROBLEMS OCCUR DURING THE LATER STAGES OF THE PROJ-

ECT. The architect or engineer will often authorize extra work or will indicate to the contractor that it will be authorized. The contractor carries out the extra work and the owners are expected to pay. This upsets the owners who feel that this extra work is related to errors and omissions on the drawings and are annoyed because they were not consulted about the extras until after the work was completed, at which stage the only recourse is to pay or arbitrate. Moreover, rather than process change orders regularly, they are held until the job is finished because of the embarrassment at having to make changes and then surprise the owners. This was the owners' most consistent complaint. One owner of a large hospital in downtown Boston said, "When I went to close out the project, I was presented with a pile of extras that had already been completed. None had been previously approved and some dated back three years. What a shock this was to the finance committee." "Why is it," they all asked, "when there is a problem the architect or engineer expects the contractor or the owner to pick up the cost of the changes? Doesn't the architect or engineer have some responsibility?"

5. ARCHITECTS DO NOT UNDERSTAND ENGINEERING. Today's building costs are reaching the point where 60% to 70% of the cost of construction is in the structural, mechanical, electrical, and plumbing systems. Yet most architects do not understand these and delegate the work to their consultants, who seldom if ever talk to each other, much less to the owner. The result is expensive systems, poor coordination, and high maintenance costs. Among the most persistent problems owners experience with their buildings is the difficulty of maintaining and operating mechanical and electrical systems in an economical way.

Naturally architects or engineers will defend their positions and counter these attacks by pointing out all the problems caused by contractors or by unreasonable or incompetent owners. The owners may not be qualified, in their view, to judge performance. This is an understandable reaction, but not one that helps the architects' or engineers' position in relation to new work. Contractors win their work by being the low bidder. They may not care what people say about them, for it is comparatively easy for them to shift the blame to the architect, engineer, or the owner.

If, for example, a contractor is to blame but is the low bidder on the next project, he or she will win the contract. The architects and engineers, on the other hand, can only win a contract by proving to new prospects that their performance for all of their past clients was impeccable. In the competitive marketing arena, it does not really matter whether the criticisms of the architect or engineer are valid. The important thing is that such criticisms are being made. It is how the architect or engineer is perceived that matters. Somewhere, the communication or service between owners, architects, and engineers has broken down, and owners are unhappy with what they have received. These owner concerns must be cleared up, either through education or improved communications, so that they no longer damage the professions. Two kinds of activity are needed: first, architects and engineers must do everything possible to improve their services and their individual reputations. Second, owners must be educated to know what is reasonable to expect and what action to take if they do not get what they expect.

This discussion brings up another area of confusion. I talked to many architects and other professionals during the research for this book. Most agreed with me that the foregoing problems are those that they are also hearing about from local owners. A typical architect's comment would run something like this: "One of our local owners just completed a project and experienced almost all the problems you noted. But that is because they did not use our firm. Yet the worst part is they have just signed another contract with the same architects to do a new building. What is wrong with today's owners that they go back for more after a firm just messed up as badly as this one did on the last project? We tried to get interviewed for the project because we do not have those problems on our projects, but the owner would not invite us for an interview. It must be a political situation, and I feel like calling the owner up to wish him well and hope that he doesn't get hurt as badly this time as he did last." From this remark, one can derive two parallel problems. First, owners are not satisfied with the services they are getting. Second, the professionals feel that although their firms are giving satisfactory service, they are not getting any work while the firms getting the work are causing all the problems. To get further information on this subject, I traveled through the southeast talking to many architects and owners. The architects continue to insist. "We do not have problems on our jobs, but the local owners keep going to out-of-area firms and, boy, do they have problems!" On the other hand, the owners I talked to were more specific about where they felt the problems lay.

Let us examine the owners' concerns more closely, realizing that understanding these concerns by architects or engineers is essential to successful communication between these professionals and their clients and thus to maintaining a steady flow of work.

The owners' biggest concern was that architects do not provide the service they sell. Before the interview, they paint a beautiful picture of the firm's performance and record of happy owners and trouble-free buildings. This impression is usually created by the chief partners, the firms' image makers. Well versed in their presentations, chief partners always know the right things to say at the right time. Their firm is hired and programming is usually completed, with no problem. The conceptual and schematic phases follow, consisting of broad-scope drawings showing the design intent. Considerable discussion occurs during these phases, but minimum work appears on paper. At this stage the partners, satisfied that the project is all set, typically turn the detailed design development over to their staff, while they run around the country looking for more work. This, say the owners, is where the project starts to fall apart. The owners feel that at this point, unless they are dealing with a very large firm, they cease to deal with a principal who is controlling the input of a given team and must themselves begin checking the work. Then as the project moves into working drawings, the owners feel that the level of professional competence is reduced still further. Nearly all owners I talked to had the standard misconception that architects take a set of plans from the previous project, erase the name, and reuse it on their project. This is a misconception that must be corrected. They also felt that many offices employ only high school talent in the working-drawing department and that very often the partners are amazed by some of the details coming from their own offices when they see them being constructed in the field. The general conception of many owners is that, after the design is completed, the rest of the process just happens. Architects no longer lead the process, but let it take its own course, doing everything in their power to talk their way out of problems, when things go wrong.

The construction phase is, in general, an unfortunate time. Most owners complain they seldom see the architects or the engineers. When they do, it is usually after the fact, when it is too late to catch a problem or make changes and the only option left is to work out the best solution to an unsatisfactory situation. Working out of the problem usually involves the owner in extra expense or in learning to live with an unacceptable alternative. In general, owners I talked to felt that architects were weak in field inspection and in project leadership when it came to making the contractor perform. When the chips were down, the architects often disappear.

Owners were concerned about failure to meet schedules. It seemed to them that materials were delivered to them on the very day that approval was needed, giving them no time for review before the request for approval was made. Another common concern was the difficulty of getting a project completed on time with all the loose ends cleared up and without litigation. Owners repeatedly said, "It seems that to get a project finished we have to be tough and threaten the architect, engineers, and the contractor." The last 10% of the project seems to be the hardest to complete and will often take one or more years after occupancy before the owners have everything as they want it. Owners say that if they waited for everything to be perfect, they would never occupy the building. A persistent cry on the part of owners is, "How can we find a way to get our projects built within the budget?"

The following is a list of owners' secondary concerns. Although these are common statements and concerns we have all heard an owner express, they should be approached with a sense of proportion, for when they are all listed at once, as here, their cumulative effect is overwhelming to the point of suggesting a lack of credibility in the architect-engineer/owner relationship.

1. TOO MANY DO NOT KNOW THE CODES. This, I think, is an unfair statement, for architects and engineers are generally preoccupied with meeting and following codes. And indeed, it becomes more and more difficult to comply with the codes that may be applicable to a single project. Making the right decisions on 99% of the code requirements does not prevent the local inspector from denying occupancy if you miss one code, and this is the one the owner is going to remember. As a result, the owner may well be heard to comment: "The building will be a disaster because the architect goofed." Although to the architect this may seem an unreasonable complaint, to the owner it does not seem unreasonable to expect 100% performance.

2. THEY DO NOT KNOW OPERATING COSTS. A typical owner comment is "Why can't architects use simple things and standard parts? A big problem is lighting fixtures. Architects pick the fancy ones that cost more money and whose replacement parts are nearly impossible to obtain for normal maintenance. The mechanical rooms are usually an afterthought and once all the equipment is jammed in by the consultants, it is almost impossible for the maintenance crew to get in the room, much less service it. The engineers must realize that the person who is going to maintain the equipment will probably be only part-time and lack maintenance experience. Yet the systems were designed by an engineer and almost take an engineering degree to operate and maintain. If we have several buildings in our system, why can't the architect and engineers accept that we have found certain items costly to maintain, and avoid them in their designs." One owner told me he had set up a book of standard equipment and approved manufacturers for the school district to try to reduce the maintenance costs and minimize the number and types of spare parts necessary to keep the buildings in operation. In spite of this, the architects who designed his buildings refused to use this guide.

3. SIMPLER SOLUTIONS NEEDED IN DETAILS AND DESIGNS. The tendency is to include too detailed, overdesigned items for the sake of aesthetics. This complaint is enhanced when the building is completed and you look at it in three dimensions instead of the two shown on paper. This overdesign is one of the reasons professionals are frequently accused of building monuments to themselves. Their buildings are expensive to build, and undue time and attention are devoted to the exterior design in hope that the building will become famous. Unfortunately, the interior often fails to function effectively; the roof leaks and the building is over the budget for both construction and maintenance. Owners are usually extremely excited when first presented with a rendering, but when the scaffolding is taken down and they take that first full look at what they bought, they may compare the finished result with that first picture. Often the building comes out better than the rendering, but sometimes it hardly resembles the rendering at all. It may even be downright ugly. But it is finished, a lot of money has been spent on it, what can the owner do but accept it?

4. NO STRONG LEADERS. As pointed out earlier, owners feel that the first phases of a project will run fairly smoothly, but that as soon as the going gets tough, architects or engineers slow down and sit back and wait. They seem unable to pinpoint what needs to be done and the appropriate action to achieve the required results. They quickly become followers instead of leaders. There is always a reason why something cannot be done. As the project moves into construction, this problem increases. The contractors take over the lead and run the show, either to their or the owners' advantage. The architects or engineers often seem to be along just for the ride, tending to ignore problems in the field; either they do not notice the problems at all, or if they do they hope they will go away.

The Response

In trying to draw attention to the common concerns of today's owners, I have not come up with any surprises. But when the concerns are brought together in one list, they are depressing. Very few owners I talked to were satisfied with their architects and engineers or would recommend a good firm that "would get your building built without major problems." Talking to architects in the same area showed me the other side of the picture. They all felt they were experts in their profession and did not make mistakes, but could not get any work because the out-of-state or out-of-area firms were coming in and getting it all. While I am not trying to cast architects in a bad light or say they should not be in business, one cannot avoid the conclusion that owners generally lack confidence in architects. Often the architects' confidence in their own performance and client satisfaction do not correlate with the story told by the owners. The burden is on the architects to take these criticisms seriously and correct the resulting misconception. But this would only be done if the architect sees the owner as a vital, intelligent team member who recognizes and deserves professional performance.

In 1970, shortly after my first experience with the owners in the New England Construction Education Program—when I found out that owners and contractors were unhappy with architects and engineers—my firm started prospecting and servicing our projects in New England. We directed our efforts to the correction and elimination of these owner concerns. We listened to what they wanted, and as a result achieved a flexibly designed building that was of good quality, inexpensive to build and maintain and fast to construct. In the following five years, our New England operation grew and expanded, while the rest of the firms in our area were reducing staff. In Maine in particular, a healthy work load was the direct result of one owner's satisfaction with our service which led him to recommend our firm to his colleagues.

In your own area, it is worthwhile studying which architects, engineers, and contractors are busy and visit some of the owners to find out for yourself what they want. But do not be defensive if they criticize the professions. Whether the owners are right or wrong their perceptions are important.

When asked to put aside criticism and state in positive terms what they wanted from an architect or engineer, the owners' first requirement was service throughout the project, combined with the ability to anticipate needs and the initiative to meet them. They wanted a firm with good experience in building types they were considering and with a highly qualified in-house staff with specialized technical and professional expertise in all aspects of building technology. An in-house engineering staff was looked on as a great advantage. Owners have a predisposition toward local firms, but do not regard this as important as experience and service record. They looked for firms who could produce a list of truly satisfied clients from past jobs.

Nearly all the owners agreed that if they were satisfied with a professional firm, they would try to give it as much repeat work as possible because it is easier to work with someone you know than to start all over again. But most of them admitted that they seldom actually gave repeat contracts, for by the time their project was over, the members of the team were no longer working together.

New Methods of Owner Protection

Evidence of how strongly owners feel about the inadequacy of architects is to be found in rather alarming measures if you look at what today's owners are doing to protect themselves. Owners with whom we contract to provide professional service are developing techniques to protect themselves from us, the architects and engineers. The trend is an ominous one.

METHODS FOR OWNER PROTECTION

1. CONTRACTS. Large owners and continuing owners are developing their own contracts for architectural service. The feeling among them is that the standard American Institute of Architects (AIA) contract protects architects and engineers more than it does owners. In their own contracts, owners try to tie architects to performance. They include a detailed time schedule to which architects must hold. They include a fixed budget that the architect must not exceed, and they attempt to include stiff penalties for nonperformance. In general, these contracts are thick, unwieldly documents that include protection for owners on every item from which they have suffered in the past. The legal aspect of these contracts is sometimes questionable, but the intent is clearly to protect the owners. More and more such contracts are being negotiated with fixed fees. The owners feel this keeps architects from overdesigning, thus raising the construction cost and their fees.

2. GROWTH OF CONSTRUCTION MANAGEMENT. This takes many forms, ranging from the hiring of a full-service construction management firm to run the project, to owners hiring a part-time staff architect, engineer, or other person with a strong construction background. When hired on staff, such a person is not the typical clerk of the works who is there to observe, but rather someone with full authority for getting the project built and to make sure all the team members perform correctly and on schedule. In the county school districts in Florida, for example, such people are being put on staff, or an architectural firm is kept on a retainer to advise the school district on construction and to make sure they get what they want. In general, this must be seen as an effort by owners to watch the architects, engineers and contractors and make sure they perform on time, within budget, and deliver the product that they want.

3. OWNERS' STANDARD DETAILS, MATERIALS, AND EQUIPMENT GUIDES. These are being set up by the larger building owners and those with continuing building programs to reduce maintenance costs by compelling the architect to use materials that are easy to maintain and whose durability has been proven. One owner told me he had a drinking fountain detail that worked successfully in one of his buildings. It met all the codes, including use by the handicapped, was a simple detailed installation, looked good, was the product of a manufacturer whose goods seldom required servicing and was easy to service when necessary. This owner further stated that he had written into the architect's contract that this detail and others of which he had similar experience must be followed. The owner even gave the architect a handbook of the details he wanted followed. "Why is it," he asked me, "when I go to visit the projects under construction, I find the details are not being followed and were not included in the working drawings?" The architect's response to the question was that they felt theirs was a better way. The owner's rejoiner to the architect was, "If you thought it was a better way, why didn't you let me know so we could discuss it before you took it upon yourself to change the detail?" In the end, it will probably cost the owner money to change the detail to what he wanted.

This owner is becoming so concerned that he is building a staff of qualified checkers for the new buildings, whose responsibility is to check every word and detail. His final statement was, "If I know from my records what buildings, equipment, and materials are easy to maintain and give the maximum life and performance, why won't an architect accept this knowledge and these limitations and create a pleasing environment with this vocabulary? Why does the architect answer, 'To insure a pleasing and functional building you must give me a free hand and unlimited budget.' Even with this freedom, I question their ability to carry out their promise, for we have many schools in our district that were designed without these restrictions which have turned out to be disasters from the maintenance and functional standpoint, even though they were published in the design magazines."

4. RETAINAGE. Owners are becoming increasingly reluctant to reduce the retainage on the contractors when recommended by the architect.

In fact, they would prefer to increase it to give themselves a more powerful tool in forcing the contractor to finish the last 10% and clean up the punchlist. They are also considering setting up a similar retainage system for architects and engineers to insure that they clean up the last part of the job and compel contractors to get the building to function as it was designed to do.

SUCCESSFUL DESIGN PROFESSIONALS OF THE FUTURE

The New Master Builder

How can your firm develop an affirmative action program to attack these needs and change the owners' perception of the architectural and engineering professions to a positive assessment? First, I feel that we must return to the concept of the master builder. This person must be the real leader of the team, in actions as well as words. As the leader, the master builder must know all the positions that need to be filled for each project, and then see to it that they are filled with qualified people. We are all too quick to say that because architects or engineers do not have any capital invested in the project, they must yield the leadership role to the contractor, developer, or owner. But this does not have to be, for architects or engineers are making investments of their time, talent, and office expense in every project and must make a profit. If the project does not get built, they will not break even or make a profit.

Next to leadership, the master builder must develop a new appreciation and understanding of the technical content of buildings in today's marketplace. Owners and contractors want ideas that can be translated into workable solutions, and they want the architect to be responsible both for the idea and its realization. With 60% of the building costs devoted to engineering functions, it is critical that the architect be fully competent in building technology.

Moreover, the master builder must be flexible enough to respond creatively to widely varying and specific needs. The temptation to rely on preconceived notions of the kinds of service, contract, or process the owners want must be resisted. The master builder must respond instead to the owners' actual requirements. As professionals, we are too eager to take typical owners and their programs and force them into the confines of a standard contract and scope of work. This is no longer a valid approach and will not work. Along with this creative repackaging of service, persons acting as master builders must be careful to promise only what their firms can deliver and not what they would like them to be able to provide. Joint ventures with other competent firms can sometimes solve this problem.

In today's market, the complexity of keeping a project going with everyone's full participation requires astute business management. This requires someone who understands how things happen and why and knows how to attract and use the right people to do what is necessary for the good of the project. Costs and schedule are two key words on the lips of all owners. And the master builder, if truly the team leader, must appreciate and respect these two factors of business success. However, the architectual and engineering professions have reputations that are shaky in these two areas. Until their mastery of these elements has been demonstrated, they will not be accepted as the leaders of construction teams.

Another way of emphasizing this point is to stress that about 90% of many architects' and engineers' time is spent in problem solving, getting people to work together, while only approximately 10% is spent designing. This is not to devalue design, but to point out that in today's competitive market, business and management qualifications are stronger factors in the marketing of services.

A typical team involved with the creation of a building will have the contractor whose strengths are a detailed knowledge of estimating and scheduling and whose key to business is bidding right so that a profit is made and the developer, whose forte is a strong background in finance and business development, construction laws, and codes. A developer is typically a professional salesperson and a doer. The owners are usually business people who have owned and operated their business with sufficient success to have expanded their company to the point where they need new buildings. In such companies, the typical architect or engineer of today is likely to be submerged by the group, thus becoming a follower rather than a leader.

Master builders can be strong, positive leaders who get things done. They will fully understand the operations of the other team members, not just their professional jargon, so that their discussions with them have substantive rather than rhetorical content. When these types of people enter the picture, they will naturally gravitate to a leadership role, and others will look to them for guidance.

An objection to architects or other professionals as team leader may be made on the basis of financial control. If they are not the ones with the money and are not paying the bills, how can they exercise effective leadership? Developers and contractors, however, as well as owners, will always respect a professional who has knowledge in their own area. They are looking for ideas and for someone to help call the shots and if the architects qualify, they are the natural persons to look to. If not, it will be someone else. The master builders' job now becomes one of strong leadership to get the people together and keep them working as a team to bring the project to successful completion. They will accept the owners as intelligent, valid team members as well as someone from whom they can learn. They will involve the owners in all decision making and problem solving as such is needed rather than after the fact. They will listen to the owners' requests, include them in the building where possible, or give a rational explanation for rejecting such. The contractor is also to be regarded as an essential team member, whose experience and expertise are important at many levels.

Becoming the Master Builder

In building up an office that can attract and keep good clients, while continuing to expand its work load in new areas, there are several areas of special concern. First, you should take the different building types and develop your background so that you are at home and proficient in your knowledge of what goes on within them. To be truly successful in school design and construction, for example, one should learn to teach. This will allow for a better understanding of what will go on in the spaces. If teaching is understood from practical experience, architects will be better designers of teaching space. They will be able to talk like educators. This will do much in establishing credibility with the staff and owner and will reduce problems in communications through the use of a shared vocabulary. Another good way to learn about the spaces that have been created is to go one step further than final inspection and acceptance. Help the owner move into the space and set it up. This will show you, for example, why doors should be large and whether the space is big enough for the programmed furniture.

Once the staff is moved in, take time to work with them to help them understand the space and how it should be used. This is excellent information to feed back into the job process and provides good guidelines for future jobs. Your additional interest in getting owners into their buildings will long be remembered by them. Do the same also with the custodians and maintenance people, for this is where you will really find out what kind of buildings you have designed and committed owners to live with. You will know the answer to the question, "Does the building work?" If you are in a school building at noon, it is a good idea to eat in the school cafeteria. If you do this often, it becomes obvious that too many architects design school cafeterias based on what looks good on paper, not on how they will really function. You should know as much as the owners about the activities the building is designed to house.

Become active in your own area of expertise. Read as much as you can about the theory behind the concepts used today. Attend the trade shows and become familiar with the latest product lines. It is embarrassing when owners know more about what is available than do their architects or engineers. Stay on top of new trends, products, and ideas.

Talk to people as you travel and visit other projects as well as your own. Find out what is happening around you and why. Learn to ask questions and then learn to listen to the answer. Gather all the information and try to reach a conclusion about what the trend is, what is happening, and how you can become involved. Then use the information to obtain more work or to perfect your service or product. Most of us talk too much. You learn by listening.

Get your office organized and set up so that the routine becomes relatively effortless and you and your staff are not spending time and money reinventing the wheel. If your procedures work well do not keep changing them. Look for unnecessary costs in all phases of your in-house work. Reduce costs to a minimum without reducing quality and service. This is vital, for you will need what you can save in overhead to devote to prospecting and improving client service. Understand the details of your own building, especially engineering and maintenance cost.

These factors may seem obvious, and busy firms are probably following these guidelines already. But if you are not busy, then something is wrong. You may think you are following these suggestions, but in reality you are not. If you try to convince yourself that all is well with your operation, you are only deluding yourself and your profit column.

Remember to establish your goal for growth. The first steps must include the following:

1. Satisfy all your clients' needs and keep all present and past clients happy.

2. Make your clients feel important.

3. Provide competent and timely services.

4. Provide creative services.

5. Prospect new work.

6. Work toward a goal in which a majority of your new work derives directly or indirectly from referrals from satisfied clients.

7. Assume a role of leadership.

CHAPTER TWO

SUCCESSFUL PROSPECTING TECHNIQUES

The process of prospecting and the development of methods for doing so are both very basic and easy to learn. The real problem is the time prospecting takes if it is done successfully. It takes constant attention and follow-up to keep the process moving.

SEEKING OUT AND DEVELOPING PROSPECTS

How to Find Prospects

In today's market, there is sufficient work within a two- or three-hour drive of any firm to keep it busy. The whole key is attitude. Architects or engineers without positive attitudes quickly slip into the ranks of the "has-been." Believe in your firm and in the marketplace. It is no good listening to other firms and contractors who cry, "There is no work." If you do, you will be sitting home feeling sorry for yourself and someone else will go out and get the work. Even though television and newspapers report that construction is down and firms are going out of business, do not adopt the doom-and-gloom syndrome. If a firm is aggressive, wants work, and has a good performance record, work is available.

In 1970, when many firms were having trouble surviving and New England was not building new projects, I opened a regional office in Vermont for my firm. Five years later, we had opened another office in Maine, and had a large work load. The New England operations grew in five years from a one-man operation to two offices with four secretaries, two construction managers, and three additional architects. Most people explained the phenomenon as unique, a case of being in the right place at the right time. But this was not so. Success derived from an attitude that stimulated us to go out and do something instead of sitting around the office telling everyone how bad it was.

Professionals should avoid the temptation to continually lunch together and commiserate with others about the economy and the decrease of new construction, for this only fosters defeatism.

When I was collecting material for this book, I visited several architectural and other offices because I wanted to get a feeling of what was happening. As I visited each office, I found that the key people were working on anything but prospecting. They all said that business was bad; they were having a problem with cash flow, and the immediate future looked bleak. They said they were trying to hang on until that big job miraculously appeared, and they were optimistic that the economy would improve eventually.

I did not believe that things could be as bad as they suggested so I dug up the name of the busiest firm in an area and went to visit its office. I was impressed. The office was so busy that everyone appeared to be working and moving on the double. Everyone who greeted me, including the secretary, had a very positive attitude. It was an exciting place to be. I was told that the office was holding up very well through the recession and would, in fact, be hiring in about three to four months.

Who was right? I decided to find the answer myself. I drove for three-and-a-half hours to another city and started working my way back. I found a large amount of new construction going on, and of course new construction generates more. I was particularly aware of the volume of new houses being built. House construction generates school problems, which lead to school construction. I stopped to see a couple of school owners. They reported that between the two districts I visited, there would be over $50 million spent in new school construction over the next five years. I found out later that all the neighboring districts expected to be in a similar situation. These were projects for which an architect had not been hired. A small portion of this action would keep any of the firms I visited busy and happy.

I then went back and talked to the "negative" firms and told them what I had found. They were not impressed. They answered, "That's about the volume that was done last year, and it all went to offices outside of the area." They thus continued to sit and feel sorry for themselves.

Your attitude can make or break you. If you sit around your office with a negative attitude, the work will not come to you. If you go out into the field with a negative attitude people will not hire you. You must develop a positive attitude and get out of the office and see what is really happening. You should not worry if the trend is down from last year. Just decide you are going to get a piece of the action. Find out what other firms are doing and profit from their mistakes. Talk, think, and act positively, and do not give way to the self-indulgence of the kind of pessimism that defeats you before you start.

Once you have the right attitude and you truly believe there is work in the area and are ready to start seeking out prospects, where should you start? The place to begin is your past clients. Visit them all to make sure they are happy with your work. If they are not, do whatever is necessary to make them so. A satisfied client is the best recommendation you can have, and a good reference for new prospects to call or visit. Let your clients know you are looking for new work and continue to sell them on the value of your services. Your clients should know almost as much about your firm as you do.

We tend to forget that a building owner is often a good prospect for another building, a renovation, or an addition. If most of your new projects are not for repeat clients, you are doing something wrong. Such repeat work can probably be obtained without an interview and with an easily negotiated contract.

Past clients must be kept happy, and they will then be strong supporters of your firm. They will provide you with referrals, leads, additional work, and will enhance your image by showing off their successful projects at every possible opportunity. Construction is a very exciting process to the typical owner, who is intrigued by the translation of ideas into reality. You must make the whole process exciting—not only in the design stage, but right through to the move-in, final acceptance, and guarantee period.

Newspapers are another source of leads. The only problem is that by the time a specific project is in the paper, the architect has usually been selected. Whenever you see news articles of pending constructions, check them out. If the projects have already been assigned, ask the owners if there is any new work pending for them or within their area.

Having tracked down obvious leads, you should read between the lines for future projects that may not be as clearly defined. You will then be prospecting for a client, before that person even realizes there is a project. For example, look for articles dealing with committees formed to study school overcrowding. This indicates a problem and will give you names of the key people to contact for further information on their building problems.

Watch for announcements about owners who have purchased land, since this usually means they will be building in the near future. On one of my first trips to Maine, after landing at a little country airport, as I was driving to an interview, I stopped at the first store I came to and picked up a local newspaper for a dime. The lead article was about a local bank in Bangor, Maine, buying a piece of property. I followed up this lead, and my firm was eventually selected to build the bank's new home office on the property I had read about in the newspaper. All this for a dime, some sales imagination, and follow-up.

Watch for news items that describe an expected influx of people because a new employer is moving into the area. You may not get the job of designing the factory, but increased population means the construction of other buildings, such as schools, churches, and stores. Use your imagination to visualize a construction project out of every newspaper article you read, then find a client and sell him your idea.

Monthly reporting service publications are subscribed to by many architects and engineers. These publications usually list all jobs in progress and pending selections of architects and engineers. Once their name is published, the owners who have not hired architects or engineers will be inundated with calls from architects within hours of the papers' distribution. This is why most owners will keep their projects out of the papers as long as possible. It is the only way to avoid being besieged by professionals looking for work.

When you find a project that interests you in a paper first check to determine if it is indeed a valid prospect, then evaluate the possibilities of an early approach to the owner to outmaneuver your competition.

Looking for leads is a process that must go on all day everyday. A distinction needs to be made between looking for leads and selling yourself and your service. In selling yourself and your service, you are constantly talking about yourself and your firm, but resist the temptation to bore people with a catalog of your virtues. There is a time and place for selling yourself and your service, but it is not with your daily contacts who will judge you by performance, not salesmanship. Daily contacts are excellent sources of leads, if you do not oversell yourself.

You can get leads from your daily contacts by keeping a low profile and learning to lead the conversation into areas you want to explore for information.

Each time you make a prospecting call, ask the person you talk to if there is anything else going on in his area that might require an architect or engineer. Owners talk to each other and keep posted on what each is doing. If you get a lead in this way, follow it up as soon as possible so that, if it comes to anything, the owner will remember you and contact you about it.

Keep in touch with local businessmen in the area of your practice, because they are in a good position to keep you posted on who might be in the process of hiring an architect or engineer. They may not have plans for their own companies, but these men are on the boards of banks, schools, hospitals, and other institutions in the community, which is fertile ground for information on new construction.

Contractors are another key source of information. They know who is planning to build. Today, prospective clients often go first to the contractor's office when they are planning to build, rather than go to an architect or other professional. They feel the contractor knows costs. Whether they are right in this, we must accept the fact that this is what they do. If you can establish a relationship with one or two of the better contractors in your area they may refer work to you and possibly even "joint-venture" with you on some of their bigger jobs.

When talking to contractors, do not be afraid to let them know you are actively looking for work. The contractor on a project might overhear the owners talking about another project for the future. This is important information for you.

Take a contractor or salesperson to lunch; it will be appreciated. At the lunch, explain that you are looking for work in a certain area and listen for his response. Each time I did this, I came out of the luncheon with a new prospect that I might otherwise not have heard about. Obviously this is not something you should do every day, but try it every month or so; you will be surprised at the information you derive.

State agencies are another source of good information. As we see many states heading for increased reimbursement of school construction costs, we are going to see them establishing priority lists. To do this, they survey all the buildings in the state and rate them according to need. The same applies to hospitals and nursing homes. This list is required by law to be published for hearings and approvals. By obtaining this list you can develop your own list of prospects needing architectural and engineering services in the near future. Since this is an obvious method of getting information, you can assume that other firms will also be using this method. The important thing is to do it first.

You can do this by developing a positive working relationship with the key person in each construction-related government agency. If you are constantly in touch with these people as part of the process of doing business, you will soon find they are involved in many other projects throughout the state. As you gain their friendship, you will find they will tell you about their work as you tell them about your other projects. This exchange of information will often give you other leads for future work. If you give away information about what you do and how you do it, so-called trade secrets, you will be surprised how much the other person will tell you. These agencies are often the first place an owner goes to find out how to go about getting started in a building program. You must have access to this information.

Your employees are another source of knowledge about what is happening in the areas where they live. This is especially vital to the small firm looking for smaller residential and commercial work. Explain to your employees that the firm needs work to survive and for them to keep their jobs and that their help is needed in finding this work. Be sure they each understand your firm's capabilities and potential. Give them brochures to distribute, and teach them how to listen to conversations that might reveal new construction. They will bring you a lot of ridiculous ideas, but treat them seriously. One day they may bring you a winner. Do not turn your back on them after they start trying to bring in new business; make them feel important. If one of your employee's ideas turn into a real prospect, be sure to keep him posted on its status on a regular basis. If the contract is signed, reward him with praise, a bonus, or other appropriate recognition. Your employees can be additional eyes and ears for you. This is important to office survival.

As you travel around the area where you practice, make yourself aware of what is going on there. A survey crew in the field indicates someone is planning to build. This could be a project for some firm. If this is a preliminary survey, perhaps the designer has not been hired. If you see a shopping center under construction, there will be stores in it needing interior design service. Go after them, if you do this type of work.

Keep an eye on the major institutions in your area. Look for overcrowding or substandard quarters. As soon as you see the need, start making your contacts. If they do not understand the start-up process, help them get organized.

As you drive through neighboring towns, keep an eye open for construction of buildings for companies representing major corporations. If they are building in nearby towns, perhaps they are planning to build in your area in the near future. Make a contact and find out their plans.

Keep track of your local businesses. They are always in need of some kind of service. A friend of mine was just starting a mechanical contracting firm in the early 1970s, when everyone said there was no work. He had considerable imagination and could see things that needed to be done. One hot day in July, he was shopping in a small drugstore on the main street. He noticed the air conditioning was not operating properly. He talked to the owner for a few minutes and then called his office for a mechanic. A short time later the unit was back in operation. The owner was pleased because he had tried to get a repair firm to come for several weeks, but they considered his job too small, and he felt they were not interested in his business. This touched a chord with my friend. If his competition thought this job was too small, there must be other "too small jobs available." He then started a door-to-door check of the local businesses on the main street. He found most of the air conditioning units malfunctioning in some way. As a result of this new work and his continuing creative thinking, which brought in more new work, he grew to become the major mechanical contractor in the state. This is the type of imagination we must all use constantly to find new work and to grow.

Be aware of population trends in your area. Is the population growing or declining? If your area is growing, this means eventually new schools, churches, hospitals, shopping centers, and homes will have to be built. If, on the other hand, people are moving out, this does not necessarily mean less need for architectural and engineering services. There may not be new construction, but there will be considerable need for renovation of existing facilities. If you have a small firm, team up with a small mechanical and electrical firm and go after this work.

Start using "cold calls" as a method of making contacts. When you have done all your research and have many names or places to contact in your search for work, pick a name that looks promising and be there at 8:00 A.M. Have a prepared outline of your firm's services and of how you can be of use to the prospective client. Get the person talking about his or her needs. If the person does not have any work or already has a firm under contract, ask for leads on someone else who might need your service. Most people will give this type of information freely.

Once you receive these leads you should follow them up. If you find it was a false lead, do not be discouraged. Ask for a further recommendation, and keep going. This is why I suggest starting at 8:00 A.M. so that you have a full day to devote to following leads. This is very interesting work that will get you acquainted with your area. You will also make many new acquaintances. You should keep a list of them for future contacts. You never know when you will use the list, but when you need a name, it will be there.

As you visit these people, listen to what they say about the kinds of services they need. Is there a service you can provide? Do not limit yourself to services you currently provide, but use your imagination. For example, everyone today is concerned with energy conservation and saving money on fuel and electricity. This could be a service in evaluating buildings and making recommendations to reduce their energy usage. Some firms have even specialized in solving roofing problems. Others specialize in correcting construction defects. Use your imagination and repackage your service for today's market.

As you determine the new need and how it fits your service, develop your total package. Start with the feasibility studies and go through construction and job training. Package it so you can sell any or all of it. When you develop your total service package, make certain it includes some service unique to your firm and which will appeal to the client. If you capitalize on this, it will give you an advantage over the competition. Use your imagination to create a project where none seems to exist and you may find a new specialty for your firm.

If you specialize largely in house design, one of the key people to know is the local mortgage banker. When you talk to a prospective client who would like to build, but cannot obtain the money, you can then arrange an introduction to your banker and in that way very likely get a job started.

Initiate the project. Be there before owners know they need you. You can help them get organized and write their requirements to best fit your service, while at the same time building up their confidence in you. If you do this you effectively eliminate all the competition.

Is a Suspect Really a Prospect?

To find the answer to this question you must know what it is you are chasing and what your chances are of being awarded the contract. Chasing false leads costs time and money. The more you know about a lead, the more effective you will be in the follow-up and the more jobs you will get.

Not long ago I was a speaker at a contractors' seminar on the topic "How to Market Your Services." One of the other speakers represented a major national home prefabrication company. He told of his technique of playing "20 questions" with a salesperson to determine if the lead suspect was a good prospect or potentially a dead end. When his salesperson came to him with a lead and wanted help on follow-up, he would not help until the salesperson could answer "20 questions" about the project. If the salesperson was unable to answer all twenty questions, then that person really did not know what he or she was going after. The questions helped determine not only if the prospect was a good one, but also what kind of prospect it was. Based on this information the manager could then plan the details of the follow-up and the closing.

There are many questions you might use for your particular firm. The list that follows can be used to help you derive the maximum background information about a prospective job. It can be modified as needed and may also be the basis of a standard office form.

20 QUESTIONS

1. WHAT IS THE PROJECT? Is it a school, hospital, office building? Is it a specialized facility with a very specific set of needs?

2. WHAT SIZE IS THE PROJECT, IN DOLLARS AND SQUARE FEET? Is it new, an addition, or renovation?

3. WHO IS THE OWNER AND WHO ARE THE DECISION MAKERS? Try to deal with the right person right from the beginning. In prospecting, it is easy to contact the superintendents of schools, since they are paid officials and are usually available. Although they probably have no vote at a board meeting, they can often veto a vote. Find out who the voting board members are. Who is the chairperson?

4. WHAT GOVERNMENT AGENCY CONTROLS THE BOARD? This is especially important since more and more projects become eligible for some form of public funding.

5. DO THEY HAVE ARCHITECTS AND ENGINEERS? This must be determined early to avoid the problem of putting a lot of time into prospecting a project for which an architect or engineer has already been hired. It's unethical and a waste of your time and money to try to replace a firm that has already been engaged and will cause a great many problems.

6. DO THEY HAVE THE LAND? If they do, it is a good indication that the project is going to move ahead to completion. If they do not, you can perhaps, provide a preliminary site evaluation and get a foot in the door. If they do not have the land, you have to assess if this means the project will never get off the ground. The size of the site will give you an indication for the scale of the project. If they have land, be sure to visit the site to become sufficiently familiar with the terrain to discuss it intelligently with the prospective client.

7. DO THEY HAVE FINANCING? If they do, the project is sure to get started. If they do not, find out the reasons and assess their ultimate chances of getting the money? Can you help them arrange financing. If so, this may be another preliminary service for your firm to provide.

8. WHAT IS THEIR TIMETABLE? Are they in a hurry or is this a long-range prospect? Have they kept their time schedule in the past? If not, why? If their time schedule is one that you cannot comfortably handle, attempt to reschedule your services to accommodate their schedule.

9. HOW DID THEY HEAR OF YOU? If this is a recommendation from one of your past clients, your credibility has already been established. If this is a new lead from a cold call you will have to start from the beginning.

10. HOW MUCH DO THEY KNOW ABOUT YOU? When you can answer this, you will know what you need to do to sell yourself.

11. WHAT IS THEIR REAL NEED? Often a prospective client will start with ambitious dreams when, in fact, his project is quite a small one. A new building might be the board's dream, although all they really need are additions and/or alterations to the existing building. If you perceive a possible conflict between wants and needs, this is another opportunity for preliminary services in the form of a feasibility study.

12. WHO ARE THE DECISION MAKERS? This is important if you are to survive the interview and presentation to the board.

13. WHAT ARE THE DESIGN IDEAS OF THE DECISION MAKERS? Ask a lot of questions early in the prospecting process to enable you to determine what kind of building will appeal to them. Are they traditional or modern? Do they like open-space concepts? Are they highly structured and departmentalized? What are their color preferences? What design ideas or materials do they *not* like?

14. WHAT BUILDING FORM ARE THEY LOOKING FOR? Everyone cares what a building looks like. Try to find out what appeals to them so that you can show them examples that will excite them. Stress your concern for energy control, low maintenance, and simplicity.

15. HAVE THEY TALKED TO OTHER ARCHITECTS OR ENGINEERS? This will give you an indication of the competition, if there is any. If yours is the only firm, you then can build your presentation as you wish. If other firms have been contacted you must build it from a competitive point of view. Be careful not to tear down the ideas or reputation of the competition, but stress your strength in areas where you know other firms are weak.

16. HOW REALISTIC IS THE PROJECT? Your evaluation of this point will depend on the amount of time and money you devote trying to win the contract. As you ask yourself this question, be candid. If you are following your hopes rather than the facts, go slowly and watch your overhead. Many a project is pursued for a long, costly period of time before you find out that the owner is a talker, not a doer. On the other hand, do not give up too soon and give the project to your competition. Questions 17, 18, 19, and 20 should be those you feel are specifically important for your proper assessment of the project. Knowing which questions to ask is important, and it is imperative that you *know* the answers.

Once you have answered these questions and have listed all the background information, approach the project by addressing yourself to the owner and his or her needs and help to get the project started. Many owners are nervous about today's market and will be slow in making decisions. This—combined with their lack of knowledge about environmental controls, general tightening of rules and regulations, and increasing costs—makes them hesitant to start new projects. This is all the more reason for you to take a leadership position and show the owner how easy it is to get a project started. If you wait for him to make the first move, the project may never get started at all. Be the leader! Find the missing parts, put the project together, and you may well find yourself with a valuable contact.

Keeping in Constant Touch

Keeping in constant touch with all your prospects becomes increasingly difficult as their numbers grow. You must have someone in the office to help you analyze your prospects and suggest new avenues of pursuit. Prospecting can be a long, lonely and discouraging process and it helps to have a colleague or teammate back in the office with whom you can talk about your progress and discuss problems. This person—because he or she knows all the details of the prospects you are after—can analyze your findings objectively and advise you on how to proceed. This person can also give you encouragement when you are tired and push you back on the road again when you are ready to give up. This teammate is a key person to your survival and to your success. Pick someone whom you respect. If possible, the person should be above you in rank in order to push you into action. You must be able to take criticism from him or her. The most successful partnership or teams are those in which the members are completely frank with each other while maintaining a positive and friendly relationship.

After you have selected your teammate, develop a list of all your prospects. List the name of the prospect, what the project is, the name of your key contact, and the schedule for selection. This list should include all valid prospects who will select architects and engineers within the next six months. Develop a second list of long-range prospects. This list would contain the same information as the first, except that selection is not expected for at least six months. You and your teammate should both have copies of each list. The original lists should be filed with your secretary, who will update them as action takes place.

Select a time when you can talk to your colleague each day to discuss progress. Prospecting is most important when you are busiest, because it takes so long to develop a prospect and convert it into a contract. When your work load slackens, and you have time to prospect, it is too late. You will use up your profit margin before you land your next job. Set aside time each day to work on new business development. It will take a minimum of one full day of your time each week and two hours each of the other days to do a successful job.

Keep a complete record of information about each prospect you have developed. This should include the background of the project, what and where it is. The names and addresses of all the people involved and a broad-scope report of what you have done in trying to land a contract. This file is essential when you come to prepare for the actual interview. It will also give you material to review when you analyze the prospect in your planning sessions. Once the prospect has signed a contract or abandons the project, the file should be placed among the inactive files for future reference, for use in the event you develop another prospect with this same owner or of a similar nature.

While we all hate to generate paperwork, we cannot survive without it. How many times have you heard someone say, "I wish I had written that down. I can't remember exactly what happened." To make the paperwork on prospecting as easy as possible, give your secretary the responsibility for keeping up the files. If you carry a pocket tape recorder with you and dictate the results of each visit as you go, at the end of the trip you can give your secretary a series of tapes to type and put into the file. While this may produce a voluminous file, it is better to have too much information than too little while a prospect is active, and it can be pruned before it goes into the inactive drawer.

Another valuable piece of paper is the prospect reporting form. This should be a simple one-paragraph statement on each prospect from the notes you taped for your secretary. This one-paragraph report would be a general statement about the prospective job and your progress. It is most important to list the names of the people you contact. This summary would then be circulated in-house to the other partners and managers. By reviewing the report, they might recognize a familiar name or situation. They could then help with a contact or with advice on handling the situation.

Control the circulation of this and all prospecting information, even among members of your firm. The war time slogan, "loose lips sink ships," applies to job prospecting. A careless remark by one of your draftsmen to someone in another office could quickly alert the competition to a prospect you thought was exclusively yours.

Another way to keep track of your prospects is to map each one, along with your past and present clients, on a large-scale map of your area of practice, mounted on a styrofoam board easily accessible from your desk. With color-coded pins, the essential information on each job prospect can be recorded and kept constantly in view. If your teammate has a duplicate of this map in his office, this will be an incentive to both of you to make those telephone calls and work to turn all the prospects into contracts.

Make regular monthly calls to past clients and make a call of a general nature every week to each of your present clients. This should happen even though a project might be in a hold status. This regular contact will make them feel important and remind them that you are working for them. Clients have often said to me, "The architect was in my office constantly until we hired him. Then he took our program and went home. The next time we saw him, he had the building all designed." Let owners know what you are doing and involve them in the decision making by leading them through the process. This will make for happier clients and speedier approval processes. When the inevitable problems arise, clients will be more apt to stand behind and support you if you have involved them in the decision making. Your present clients are your best references and sources of leads for other projects as well as for their own future building projects.

SELLING YOURSELF

Be Enthusiastic

Be enthusiastic as you approach prospecting. The hours will be long and the results sometimes discouraging, but you must learn to never let it show. You should create an aura of excitement in everything you do and wherever you go. When you walk into an office, things should spring to life.

Your personal image is vital to your success. Learn to sell yourself. Prospective clients will buy your product only if they are impressed with you, and this will depend on how you conduct yourself. Your firm can be the best in the nation, its work published in all the national magazines and attracting phone calls from many, but if its representatives put owners off when they meet and are unable to sell themselves as persons, your firm will not get the work.

I was in the process of hiring an architect to help on one of my firm's projects. After finally sorting out the many applications, we settled on two applicants to interview. I was shocked when the first man came in. He was dressed in a strange style, which I knew our clients would not accept. I tried to overlook his dress and get on with the interview, but it was difficult, since he just sat nervously in the chair, waiting for me to start. This man was a licensed architect and should have started selling himself the minute he walked in the door and met my secretary. He wanted to become a member of our firm and work with the clients, yet he was unaware of how to sell himself. He should have started the conversation by telling me his strong points and should have controlled the interview by leading me to his good qualities.

When I had determined he was not going to take the initiative, I asked him to tell me about himself. This was his golden opportunity to sell himself to me, but he answered by saying, "Well, what would you like to know?" I had already made up my mind that this was not the kind of person we wanted to represent our firm. Even though he was licensed with several years' experience, he could not talk me into hiring him, much less sell himself and our firm to existing and prospective clients.

If you learn to sell yourself, you will find life more exciting and your business more prosperous. Most people like to find out about other people's careers, and they look on architecture and engineering as exciting professions. Confirm this impression when you talk about buildings and convince them with your own enthusiasm for what you do.

There is a direct relationship between the architects or engineers interviewing for a job with a firm and a project manager interviewing for a design contract. The typical owner who is going to build is besieged by architects and engineers, all requesting to be considered for the job. The most important factor in the selection is the image you make as you walk through the doorway and start the conversation. The most important and critical time is the first thirty seconds of your visit. The initial impression you make on the owner is what he or she remembers. If you do not impress him or her then, showing that you stand out from others, your chances of winning the contract will quickly fade.

A strong relationship exists between individual and firm interviewing. The only difference is that the individual being interviewed for a staff job is selling his personal experience, while the professional is trying to expand his practice, selling the experience of the firm. The firm's experience may be demonstrable, but the person making the presentation still has to sell himself as an individual. If you can sell yourself, everything else will fall into place.

One of the keys to acceptance by the people you are visiting is to always be expected! There are two schools of thought on this. Some say you should arrive unexpectedly so that you cannot be put off. If you call in advance, the owner can tell you not to come and then you have lost the chance for the contact. This approach may work for some people, but I dislike dropping in on people, just as I hate to have them drop in on me when I am busy. I feel this is impolite and disrupts the other person's work. If someone is busy and you disturb that person, it will probably be seen as an annoyance. Another consideration is that you may spend several hours waiting when you arrive unannounced. Decide if your time is better spent waiting or working. The only time I use the "cold call" approach is when I am starting to research a prospective job, but do not know where to start or who to see; or if I have tried to see someone who has put me off, then I will drop in to say hello. Such a visit should take place in the reception room and should not last more than five minutes unless the person you are visiting asks you to stay longer.

If you are busy and want a productive meeting, make an appointment. In this way you will be expected and your prospective client will know what you wish to talk to him about. This saves you both a great deal of time. Once you have an appointment, arrive a little early so that you are ready to enter the office and start your conversation at the appointed hour. Build sufficient time into your travel schedule to allow you to arrive early, at every call, taking into account possible problems and delays. Arriving on time will help you establish the reliability of your firm. Although people may not notice if you are on time, they will definitely notice if you are late. If you arrive on time and the other person is running late, the fault is that person's and may work to your advantage, since you will likely be given an apology. If you are kept waiting for more than an hour, excuse yourself with the prospective client's secretary and depart, but make an appointment for a later visit. Never show that you are upset about being kept waiting.

Always arrive with a flare of excitement. Practice entering a reception area and approaching the receptionist. There should be a spring in your step and a smile on your face as you approach the receptionist. Speak clearly and confidently so that you can be understood. Say who you are and whom you wish to see. This is not the time to be insecure.

If a meeting has been arranged for you by the existing client or mutual friend be sure to keep that contact posted about what is happening and the status of the prospect. It is a bad mistake to involve people in the prospecting process on your behalf and then ignore them. They will justifiably feel they have been used and may not be anxious to help you the next time. If they are made to feel important and a part of the team, they will continue to help you and play a vital role in your success. Keep them posted and thank them for their help.

Learn to be a positive person. Think, behave, and talk in a positive manner, and learn to be happy in every aspect of your work. Try never to have a bad word to say. If you are overheard maligning someone, you will quickly gain an unpleasant reputation, since people will feel that you are likely to do the same about them in their absence.

Learn how to express your thanks in a personal way. If someone has really helped you, let them know you appreciate it, not necessarily with an expensive token, but with a thoughtful gesture.

Early in my career, I was a clerk of the works on a large high school construction project. I wanted the project to run perfectly and be completed on time. Like most projects, it began to fall behind schedule. I started to push the workers in a personal way, realizing I did not have any authority to get them to work harder or faster. Because I took a personal interest in this project and was there everyday, they soon saw that finishing the project on time was very important to me. It soon became a joke that this project, which had been eight weeks behind, was going to be finished on schedule, the Friday before Labor Day. One morning at our coffee break, one of the workers told me it was my turn to buy the coffee and donuts. This gave me an idea. The morning coffee break is very important to the workers on a construction project. I told them if they finished on schedule, I personally would throw a party on the last day. This gave the crew something to talk about and work toward. The project was finished on schedule.

On the last afternoon in the new cafeteria we had two large cakes with the words, "Thank You" written on them and coffee and cokes. I invited all the school staff who had been involved in the project as well as the foreman and superintendents of the construction crews. The workers were the guests of honor. It made an exciting climax to a long year of construction. Instead of turning the building over in the usual manner, we provided a little excitement for the owners as they moved the new furniture into the new school.

On another occasion, I was working with a superintendent of schools, getting his building project ready for a bond vote. As usual, he wanted six months' work done in three. We told him we could do it, but that he was going to be a very busy man since he would have to be involved with the project on a daily basis and would have to make several trips to our office and other places. For the next three months, he would not be home very much. But he said, "Let's go." So we did. Toward the end of the three months, he mentioned that, as we had predicted, he was not home very much. I reported this conversation to my partner, who suggested sending a dozen roses to the superintendent's wife. These roses were well received and showed that even though we were a large, busy firm, we cared about people.

This kind of gesture can serve you well as long as you remember that an appropriate gesture need not be extravagant.

Always Be Busy, But Always Have Time

This is the mark of a successful person. It has been said that "if you want a job done, give it to a busy man." If you give someone a job to do when he is not busy, it will take him a long time to do it, but when he is busy and only has a small amount of time, he pushes himself and accomplishes a lot more. Capacity is only a state of mind. If a person wants to do something, he or she will find the time. We know that 90% of the work is done in the last 40% of the allotted time. We put off the real production effort until the deadline is near. The time it takes to do a job will expand to fill the time available. Always be busy, but always have time to make the person you are talking to feel that his or her problems are the most important and that you are eager to talk about them. While you must give people all the time they need, you must also learn to control it, how to lead a conversation, and how to end it effectively.

When you are asked how things are going, always say you are busy and that your business is expanding. Dr. Norman Vincent Peale's power of positive thinking really is important. Believing it, will help to make it happen, but it will not substitute for hard work!

As your prospecting brings you in new work, be careful not to ignore your present clients, the people who made you busy. No matter how busy or involved you become, you must continue to make everyone you talk to feel welcome and important. If you cut someone short or make that person feel unwelcome, they will not come back. Without people, you may be without problems, but you will also be without work.

Make your lunches with prospective clients meaningful visits, by discussing real issues as if you were conducting a client meeting. If you are lunching with members of a board, take them through a theoretical building program. In this way you learn each other's language and understand each other's job. This can help greatly in both getting the job and in doing the work once you have the contract.

If you are going to a meeting that a client or prospective client is also to attend, offer to drive him or her there. This is important and may be a key to strengthening the bond between you. Traveling together in a car is a good way to get to know someone, since you are, so to speak, captives of each other. When you are going to visit one of your construction sites consider inviting a prospective client to come with you. Most people appreciate an opportunity to visit a building under construction.

At one time, I frequently traveled in New England in my own plane and, whenever possible, I took one of my past or prospective clients with me to visit interesting jobs. It cost no more to fill the extra seat, and the return in goodwill and credibility was immeasurable. As you talk and travel with prospective clients, be aware of the power of the conscious and subconscious messages that go beyond the more obvious meanings of what is said.

Be especially courteous to your client's secretary. He or she is important to the client. Always treat him or her warmly and with praise. Take a few minutes to talk to the secretary, and get him or her to like you. Once there was an overload of paperwork on one of my jobs and the owner's secretary had to help produce the necessary documents. I brought the secretary flowers next time I visited my client's office. The secretary was excited, and my client was pleased. It was an inexpensive way to show that I appreciated the extra effort made for me.

Try to be flexible when a client asks you to do something that will mean a change in your own plans. You must learn to respond positively to all requests. When a prospective client asks, "Can you be here tomorrow night to meet the board?" You must say yes and make the necessary changes in your schedule to accommodate the new meeting, which might result in a new contract or continued client satisfaction. A client will only ask you to do something or be somewhere he thinks is important. You must learn how to respond affirmatively and figure out the logistics afterward. Do not involve the client in these logistics, since this might embarrass him or her for causing you trouble, and in the future make him or her hesitant to ask. Keep the channels of communication open and do not allow the client to feel that he has inconvenienced you.

Selling yourself, I repeat, is the key to your personal success and also your firm's success. In this sense, you are the firm. I have often wondered why firms spend thousands of dollars on brochures and other promotional tools while ignoring the grooming and training of the people they send out into the field.

Before you hire a secretary, tell him or her that part of the job requirement is dressing smartly and greeting people pleasantly in person and on the phone. The person will listen, because this is what he or she is being hired to do. But if you hire the person first, then lay the ground rules, you may encounter resistance.

Any employee who meets your public should be schooled in personal grooming and in the art of selling himself or herself.

Disney World in Florida is a perfect example of the successful training of its employees to sell themselves and their organization. No matter what the request, how long the line, or how late the hour, I have never met a Disney World employee who was upset or failed to be polite and helpful. Even the people who sweep the streets seem proud to be working there. If Disney can do it on that scale, surely it can work with the relatively few employees in architectural and engineering firms.

Are you someone people want to know better? Do you dress as if you know what you are doing? Do you look professional? If not, perhaps this is why you are not getting the volume of new work your firm needs to flourish. A checklist of important points in grooming and general bearing follows.

Appearance

1. HAIR. Neat, clean and well combed.

2. FACE. If you are clean shaven, carry an electric razor with you and make a point of shaving before afternoon or evening meetings. Keep skin blemishes under control. There are many types of facial creams that will minimize these problems. Take extra care with your neck and collar.

3. CLOTHES. Your clothes influence your reception. You must learn to dress for the people you are meeting and for the image you intend to project.

   I personally realized the importance of this one day in Miami when I arrived after sailing for a couple of weeks. A cashier's check had just arrived and I wanted to get it cashed before the bank closed. I washed, put on clean Bermuda shorts and a T-shirt, but when I arrived at the local bank, I was coolly received and the check was refused. They would not even call my home bank to verify the funds. They treated me like a tramp. After feeling so uncomfortable in the bank, I decided to experiment. I went back, changed into a pair of dress slacks with a shirt, tie, and jacket. I even trimmed my beard. When I walked into the bank, I looked like a professional who could very well have owned the bank instead of a local tourist who just got off the boat. The guard did not even notice me this time. I walked straight to one of the vice-president's desk and presented my check. It was cashed at once and I was on my way.

Everyone wants to be his or her own person and dress and behave as best expresses himself or herself. But when you are selling yourself to people—often on their terms—you have to conduct yourself and dress accordingly. A good clothing store can help you build a good and current wardrobe. One rule to remember is to buy clothes that travel well, both on you and in the suit case.

4. HANDS. A good handshake inspires confidence and communicates something of yourself to the other person. If your hands are dirty or your fingernails unclipped, it marks you as a nervous or careless person. The hand must, if possible, be dry. A wet hand suggests nervousness. The greater your confidence, the less likely you are to exhibit physical signs of nervousness.

5. SHOES. They also tell a story. They should be polished, good quality, and unscuffed. Good shoes will stand out in a crowd. An amusing incident illustrates this. I was working in a sporting goods store during the ski season when a man dressed very poorly and looking half-starved walked in. He looked as though he could not afford to eat, much less ski, so I took him over and started to show him the bottom line of clothing and equipment, but could not get him excited enough to buy. Soon one of the saleswomen came over and started showing him the top line of clothes and asked me to show him the top line in boots and equipment. I did so, and soon he was completely outfitted in the best. He reached into his pocket and paid his large bill with cash. I was the most surprised person in the store. I later asked the saleswoman how she knew that he would buy the best. She said, "It was simple. He had on high quality, expensive Nettleton shoes in good repair." She said that in retail sales training, they teach you to look at

a person's shoes first, before deciding what price range they can afford. The theory is, that although people may have many outfits of clothing ranging from the finest to the most casual, they seldom have more than three or four pairs of shoes. Thus shoes are a good indication of status, taste, and income.

6. PERSONAL HYGIENE. This is important and should be given strict attention at all times.

7. SMOKING. If you smoke, be careful not to offend other people. Try to avoid it in someone else's office, unless the meeting is prolonged and then ask before lighting up. Keep the smell of tobacco from your clothes and your car.

Self-Assurance

1. POSTURE. Sit and stand erect. Your body language should show that you are in command of yourself and of the situation. Assume that everyone has read the books on body language and project the signals of a leader. Always look at the person who is speaking. When someone else is doing the talking do not slide down into your chair as this can be seen as an insult, indicating your lack of interest in the remarks.

2. BEARING. Stand and walk in a confident and alert way, as though you always have somewhere to go and something to do. The exhausted or defeated person walks as if every step required a decision, with shoulders rounded and stooped from the burden of life and head lowered with downcast eyes. The expression of the face is one of far away thoughts and fantasies or of deep anxiety. Such a bearing does not instill confidence in a prospective client. The aura of success is important.

3. VOICE. You may have made a good impression with your appearance and bearing, but what happens when you begin to speak? Before you approach a client, know in advance what your opening remarks will be, but don't let them sound stilted or insincere. When it is time to say something, speak up, use a clear voice and look directly at the person to whom you are talking. Do not drop your head or your voice. Think

of the impression you get from a salesperson who enters your office and starts to introduce himself or herself and the company being represented, but halfway through becomes self-conscious, looks at the floor, and drops his voice. This lack of self-confidence does little to recommend the person to you, and the unwillingness to meet your eye suggests that he or she does not believe in his or her own message.

4. RESPONSIVENESS. Once the introductions have been made and the preliminaries are over, the pressure will begin. You do not know how or where the conversation will go. You must be prepared to think on your feet and respond to a fast-moving situation. Practice this with other managers in your firm. Have them toss questions at you and try to catch you off guard. Keep practicing and be tough with each other. The art of having responsive dialogues is a skill that can be acquired.

Gaining self-assurance and overcoming fears are of prime importance. Through constant drilling and building of experiences, one comes to realize that very few situations or individuals are completely new or unique. Rather, all are variations of past experiences with other people and in other places. Once you identify your goals, there is no need to be nervous. Other people are usually just as nervous as you are in similar situations and will understand your position. If you do find yourself getting tense during an interview, do not be afraid to say something like "Stop and let me catch my breath."

Avoid losing your temper. A calm, cool, self-assured person is the kind one wants on one's team. Ill temper shows through and makes other people ill at ease. Nervous habits that are especially annoying are the rattling of coins in the pocket, constant pacing up and down, or obvious perspiration on the hands and face. All can be controlled with practice.

How do you react under pressure? If someone says, "You are wrong," what is your response? It is always possible to hold another point of view; so do not show your anger if someone disagrees quite vehemently with you. Your facial expression can reveal anger. If you are biting your tongue or getting hot under the collar, the other person will notice it at once and feel that he or she has you cornered. If the confrontation involves a friend, you will probably be all right. But if it occurs at a public meeting and the person talking is trying to discredit you and your project and sees that you are upset, you will get hit again and again. The best way to handle this is to agree up to a point and then qualify your answer.

Never try to prove the other person wrong. If you succeed, you may make an enemy. Offer enough additional facts to show your position is also correct, but do not try to discredit him or her, especially in the presence of others. He or she will never forgive you for this and will be "out" to get you at every available future chance. If the person is a full-time board member he or she can hurt you more than you can hurt him or her.

If someone says you are wrong, and you are—meaning you or your firm did make a mistake—what do you do? The best solution is to immediately admit your mistake. Do not become defensive or angry or try to deny it. If you have a valid explanation, offer it, but do not make up a reason or an excuse. If it was an oversight, say so and take your medicine. If it was an error in the decision-making process, give the reason and accept the blame without shifting it to someone back in the office. The owners will recognize this as a weakness, and you will lose their respect. Admit your error, correct the problem, and make certain it does not happen again.

How should you respond to complaints when an owner makes a charge? Remember, his or her complaint may be valid. An owner may try to prove you wrong to make you demonstrate that you are right, but he or she will seldom accuse you of nonperformance if he or she does not have some valid complaint. Many architects are prone to be generalists who do not pay close enough attention to the details. They promise they will do something by a certain date and then inattention to details causes them to forget the commitment. To an owner, his is a major project and when you say you will do something it is an important commitment, and usually one cannot deny his complaint.

Resolve to promise only what you can deliver and then do everything you promise. Above all, do not become involved in a battle of words or in trying to shift the blame of nonperformance to someone else. If you become argumentative and try to give proof without documentation, the owner will produce his own facts and discredit you.

Learn to be relaxed at all times. A relaxed person expresses the kind of confidence that impresses the prospective client. Take all situations in your stride and learn never to lose your poise. When you encounter a problem, compare it to other problems you have had. Do not let it become a crisis. Keep your perspective. Learn to respond to problems with positive action. But do not relax to the point of appearing not to care.

Never joke about problems. The owner will not appreciate it. The problem is real to him or her and requires a solution.

Learning to be in control of all situations will establish you as a leader and someone who gets things done. If you see a negative situation developing, find a positive counter measure and initiate the action. Plan all meetings and control them to be sure you accomplish what you have set as your goal. Do not carry your personal sales techniques to an extreme or you will be accused of being dogmatic. Part of your leadership training in controlling situations is to develop the technique of getting others to do what you would like them to do, but making them think it was their idea in the first place.

CHAPTER THREE

PROSPECTING TOOLS

Getting ready to go to an interview is a very complex operation and a great deal of time and thought must go into the actual preparation. Do not waste time doing the repetitive things necessary for each interview. The materials common to every interview should have already been assembled and your time would be better spent rehearsing the presentation itself.

MATERIALS

First-Aid Kit

I found that I could save valuable time by developing a "first-aid kit" containing all the materials I had ever needed at any prospect visit or interview assembled into a special briefcase and kept by my desk at all times. If a prospective client called me at 10:00 in the morning and asked if I could come to an interview that evening, I was ready. I had all the necessary materials in the briefcase. This first-aid kit should remain intact at all times, or it will lose its effectiveness.

The following is the list of items I included in my first-aid kit. You can use it as a guide for developing your own prospecting kit.

1. STANDARD OUTLINE OF A THEORETICAL INTERVIEW. This would be a comprehensive outline to serve as a reference for the interview if you did not have sufficient time to prepare a specially focused outline for the individual client.

2. OFFICE BROCHURES. Half a dozen copies should be sufficient. Staple

your business card inside the cover, and include any special inserts that you customarily use in prospecting.

3. EXTENSION CORDS. These are essential. I would suggest you carry at least two 25-foot cords. The outlet for the projector always seems to be in the next room and an extension cord is almost impossible to find in the evening after the custodian has left. If you only have 45 minutes for the interview, it is a waste of time to spend half of it looking for an extension cord. If you are going to use slides, bring your own extension cord.

4. A PROJECTOR. Keep one ready by your desk. Make sure it is in operating condition and that someone has not borrowed a part from it. The projector is a vital tool to your existence. Learn how to operate it and to make minor repairs.

5. AN EXTRA BULB FOR THE PROJECTOR. This is essential. Bulbs burned out during two of my interviews. The first time it happened I had a replacement and was able to continue without too much trouble. However, the second time was a disaster. I had reached a high point in the slide presentation and the slides were vital to the discussion when the bulb burned out. No one from our team had a replacement nor did the owner. The interview was lost for lack of a simple bulb.

6. SET OF GENERAL SLIDES. This would be a collection of slides telling the story of your firm's work. It would not be the set you would use for a specific interview, but could be used in an emergency if you had no time to assemble a specially focused set. Keep the slide set intact, and do not borrow from it, intending to replace what you removed, since you will not have time to find the missing slides on the day of the interview. This set of slides is like a spare tire. If you let some of the air out and do not replace it, you will be left with a flat.

7. MASKING TAPE. This is a vital tool. You will often arrive with a display you want to hang. You may have brought along some thumbtacks only to find there is no tackable surface. The tape can be used for many other emergencies such as repairing torn papers and sealing packages.

8. EXTRA BUSINESS CARDS. You never know when you might need extra cards, and it is embarrassing to find that you left them on your desk or did not bring enough to go around. Since the others from your team may find themselves in the same predicament, get a box from each person who might attend interviews, and put them all in the bottom of the first-aid kit.

9. PADS OF PAPER. Several should be included. Never be in the position where you have to take notes on a paper napkin or ask a member of the client group for some paper. Paper for taking notes and organizing your thoughts is vital.

10. PENS AND PENCILS. Don't forget these. You may have one for yourself, but what about your audience? If you want them to take notes about your presentation, bring along a box of sharpened pencils for them.

11. BLANK CONTRACTS. Have these for all the negotiable alternatives. Owners will often ask about the contract and it is better to discuss this while looking at an actual contract form rather than talking in generalities. During the interviewing and sales process, the owner listens, but doesn't remember everything said. When you talk in generalities, especially about the contract, owners may remember only the parts that are to their advantage. This could hurt you when the time comes to sign the contract or to perform the service, if the owners then remind you that you made certain statements at the interview. They may have taken your words out of context, but you could lose a job or lose the owners' respect and cooperation as well as their trust.

12. COPIES OF EXISTING CONTRACTS AND PROPOSALS. These, along with short-form contracts for preliminary work, should also be included. The names of the clients can be blanked out. But leave in the type of project, fee, and scope of work. This will give the prospective client something to look at and compare with your proposal to him or her. This establishes credibility. You should not, of course, encourage anyone to actually sign a contract without first taking it to his or her attorney and obtaining that person's comments and recommenda-

tions. Architects are not in the business of preparing legal documents for contracts. You offer these only as examples of past documents for contracts, used by different clients, whose attorneys have found them satisfactory for that particular project.

13. FEE SCHEDULES. The question of fees will undoubtedly come up during the conversation. Use of the standard published fee schedules is considered illegal, but owners still refer to it, and some professionals still try to hold to the schedules. I recommend you do not use the published fee schedule. I could never understand how a group of professionals could sit around a table and derive a fee schedule that would relate to all the different sizes of firms and different types of projects. Each firm has a different overhead factor, and each project has a different scope of work. This is especially true in today's market where owners are customizing their requirements and are not following the previously accepted AIA format for scope of work. It is vital that you develop your own office fee schedule for each building type and scope of work in relation to the various options. This schedule must relate to your in-house costs and be set up to allow a profit on all projects. If you operate for very long without a profit, you will be in trouble. Even if you just break even, you expose yourself to future financial obligations. Without a margin of profit, you will not be in business long. Set up a reasonable fee schedule based on a realistic appraisal of your financial history. Do not be afraid to show this to the prospective owners. When the time is taken to explain the fee to the owners and discuss what it covers and your past and projected profit, they are less likely to question your fee. By following this procedure, I have almost always been able to obtain a fee higher than the unpublished official professional fee schedule. If, however, you find that you must receive an excessive fee to make a reasonable profit, you had better take a serious look at your total operation. There is no reason why an efficient architect cannot make an acceptable profit practicing architecture today.

14. SUPPLY OF STANDARD OFFICE FORMS. Remembering that you want to get people involved with you in thinking about the specifics of their projects, it is a good idea to have your in-house tools to help you do this. If an owner asks you if you can have him or her moved into a

building by a certain date it is easy to say yes. But if you have an office form that lists all the steps and times necessary to complete design and construction, show it to the owner, and sit down with him or her to fill in the blanks for the job. This organizes the discussions and demonstrates all the steps in the process and what happens if one stage is late. You have now done two things. First, you have shown the owner you know what you are talking about and have listed all the details. Second, you have made him or her think positively about getting started. He also will see the need to move rapidly at the beginning, or it will affect their desired occupancy date.

This type of form can also be developed and used for costs, material selections, room schedules, and site selection. But be sure you can perform within the limits you set down, for the owner will hold you to what you tell him.

15. PROSPECTING FILES. Store these near the first-aid kit so that you have easy reference to them and can get out the necessary folders to take with you. You should also have a copy of your summary prospect file available.

16. FLASHLIGHT. This is the final item I have in my prospecting kit. Everyone laughs when they discover this as they cannot figure out its use. But it has a great many uses, from changing the bulb in the projector without turning on the house lights, to being able to see to melt the ice on your car lock at midnight in the unlighted parking lot when everyone else has gone home. It also helps when you are trying to find something in the bottom of your briefcase in the middle of your slide presentation.

Handouts

Handouts are another valuable prospecting tool, and can contain many different kinds of information, from very formal professionally produced brochures and inserts to fact sheets, and Xeroxed press clippings. Each is valuable and has its own benefit. In my opinion, architects spend too much money on professional brochures and slides. Clients are often reluctant to take them because they know that they are expensive and will probably end up in the wastepaper basket. If you used some of the money spent on these high-priced aids for customized material produced in-house and for training your staff in interview techniques and how to meet the public, you would have a better return on your investment. The best sales tool is a well-trained manager who is constantly making contact with prospective clients on a one-to-one basis.

The formal brochures are usually glossy, professionally produced documents that represent a large investment of time and money and are difficult to customize or change to meet specific circumstances. The material included in them should be the standard information on the background and history of the firm and its principals. It also should include details of the type of services; in fact, it should list all nonchanging information. Generally, it is best if the material and layout can be arranged to tell a story. By this I mean you should be able to place the brochure on the desk in front of potential clients and leaf through it page by page, giving an organized presentation as you go. If you do not go through the brochure with the clients and explain the material, they may only flip through it and then flip it into the wastepaper basket.

Customized inserts are the most valuable tool you can have. These should be inexpensively produced copies of information pertinent to a specific project. Examples of these are typed lists of all the projects (with the names of the owners) your firm has completed. You can have a separate list for each building type, construction category, or location. It is also handy to have a list of all your current references listed by building type. Another valuable item is a roster of your firm's employees, giving their position and educational background and responsibilities.

Another useful tool was a copy of a case study I had prepared on a flexibly-planned school built by the "systems" method. The systems method referred to the design of a building using modular standard, off-the-shelf parts. These parts included everything from the structural and mechanical systems to partitions and cabinets. All parts were capable of being relocated by the school staff. The case study gave all the background on this new type of construction so that owners could relate it to their own projects and ask questions. Once again, the owners started creating their buildings in their minds, using my material. My material was then used as the basis for judging the ideas and methods of other architects.

You could also hand out fact sheets of two to four pages on specific projects. These would include photographs of the projects together with a floor plan, detailed cost breakdown, and other vital statistics. By putting each project on individual sheets or foldouts, you can select those appropriate to a specific prospect and avoid the resulting waste when all your examples are bound in one package or included in the brochure. If a person is interested in a hospital, you are wasting your material and money if you give him or her too many examples of schools or other buildings in which there is no interest.

Your brochure and other glossy material should be used for people and situations where they will be used constantly. Give all your clients a copy for the reception offices in their buildings. This will assure long-term use. All employees should be given a copy at an esprit de corps session to build their interest in and support of the firm. They will then take it home and show it to their family and friends, giving you additional local exposure. You should give the material to new prospects and contacts, but since you will also have a lot of other material, do not give it all to the potential client at one time. This is overpowering, and the client probably will not look at any of it. It will also leave you with nothing to give the client on follow-up visits. The best procedure is to analyze the prospecting process and establish a system of passing out your literature. Have a basic package for your first visit, then three or four related packages to serve as the reason and discussion points for your subsequent visits. If you give it all away on your first visit, you will not have an excuse for a follow-up.

Handouts are valuable because they will remain with the prospective client long after you have gone. They should be valuable to an owner entering the construction process for the first time by answering his or her basic questions about your firm, the quality of its work, scope of experience in his or her type of building, cost, and schedule performance record.

These handouts will also present the firm's image of itself, not only in the design of the building pictured, but also in the quality and content of the document itself.

While it may be necessary to have formal, professionally produced brochures and handouts, avoid the trap of putting too much money in them. Save a high proportion of your development funds for the training of your manager to meet the public, for traveling and prospecting expenses, and for taking care of your existing clients, your best source for work.

Project or Office Visits

Visits either to one of your projects or to your office can be a key selling tool or pure disaster. Timely, appropriate, and well-executed visits will leave a favorable impression of you and your firm with the prospective clients. But if you select the wrong place to visit or plan the event poorly, you will leave a bad impression, and someone else will get the contract. You must take almost as much time planning such visits as in planning interviews.

The reasons for the visit are many. Up to this point, you have been meeting with prospective clients on their grounds, in their offices where you are totally under their control. The visits are usually with only one person for a short period. Even the interview, where the entire board is assembled, is on the clients' territory. It is important for you to get them out of their territory and into yours, where you can control the situation and time, have their undivided attention, and determine what they will see and hear. You can control where they go and how much time they spend there. You are their guide, and they will see and hear only what you want them to.

Such visits will allow them to get out of the roleplaying they are required to do on home ground. This will get them talking more freely, and listening more closely to what you say. At this time, they can meet and talk to other members of your team and get a feeling of the depth of support your office provides. They can also meet your past and existing clients and get to talk to them in person.

Visits are excellent tools in allowing you to show owners who you are and how you perform. I always try to end such visits in a conference room to summarize what has been seen and discussed and work toward closing the deal. The visit should be similar to the interview in that all stops through the day should build to a high point of excitement at the end. Ideally, this high point should be followed by a meeting at which the deal is closed. Plan what you are going to see so that each item you show them is better than the last, with the best and most exciting shown at the end.

Travel arrangements and planning can make or break the visit even before the prospective clients arrive at the office or site. Decide whether they will be responsible for financing the travel arrangements or if they are expecting you to pay the bill. This should be decided early and made quite clear. If it is your standard procedure to pay for the travel, offer it at the time of the invitation. This will get it out on the table and will avoid the embarrassment of their asking you to pay the travel expenses if they do not have the funds.

Whether they pay or you pay, the travel arrangements and planning must be made and coordinated by you and your office. You have done it before and know the many things that can go wrong and the pitfalls that can develop if anything is left to chance. If you are trying to assemble five or ten people at 6:00 A.M. and transport them to a site two or three hours later, this is a difficult situation. Add the problems of arranging your travel and that of any staff you want to be there, as well as the owner of the project you want to visit, and you have a formidable task. You are involving a lot of people in this visit and are taking their time. It is frustrating to arrive at the site with your team all assembled, only to find the owners will be two hours late or have had to cancel.

Determine what you want to show the owners and where it is located. Then consider the two phases of travel involved: from their home to the site and from site to site during the day. The return home will be a repeat of the method used to get them there.

If your area of practice is generally within 100 to 150 miles of your office, your prospective clients will probably live within this radius also. This makes traveling relatively easy. Private car is the best means of transportation, and the clients will probably arrange their own transportation to the site.

Departing at a fixed time is most important. You must insist on some kind of telephone check so that when the last person has been picked up and the trip has started, someone at the location gives you or your office a call to tell you they are on their way and confirm the time they are expected to meet you and your team. Be sure they have a telephone number for your office and the name of your secretary so that if any trouble is incurred enroute, the clients can call and advise the secretary of the problem. This will allow you to keep track of them, and if something goes wrong, you can revise your plans so that when they do arrive, you can continue with your plans. You must set up the same checking system with the people from your office so that you know the status of everyone on the visit. This eliminates unnecessary worry and allows you to concentrate on the visit itself.

Visits of the kind that I was involved in were usually over 250 miles. It was necessary, therefore, to consider air travel because we wanted the people to arrive fresh and relaxed, and it was preferable if they were back home the same day. This brings up another series of problems for which you must make alternate plans. If you use commercial flights, you must be prepared for schedules that make it difficult to do a day's work and return the same day. There is the constant problem of planes running late or being diverted to another field because of weather and of service being limited to major cities. When you consider all these factors and check out the total cost of moving a group of five by commercial airline, you may want to look closely at the local air taxi operations. You can generally pick up people closer to their homes and at a time convenient to them. You can arrive at the site by 9:00 A.M. without having to get up too early in the morning. You can visit other projects until around 4:00 P.M. and still get them back for dinner. The price will probably be comparable to a commercial airline, but you will have a much more efficient travel schedule.

Become familiar with the air taxi operators in your area of practice. Know who the good ones are and who should be avoided. Most people have never flown in a small plane, so you must make sure the experience is pleasant for them. If they arrive and are apprehensive or unnerved, they will not be very interested in seeing your buildings. All they will be thinking about will be the return trip. Check out the air taxi services closely so you know who to use for the different areas of your operation.

To evaluate an air taxi operation, stop by the facilities sometime when you are prospecting the area and take a look around. See if the buildings are neat and clean and if the office is run in a professional manner. Beware if you see many people who do not work for them sitting around the area; they are probably swapping tales about their dangerous flying experiences, which will quickly unnerve the first-timer on your board if he arrives early and listens to the gossip. Make certain the people around the hangar are professional and talk business only. Next take a look at the planes they use. Are they clean inside and out? Do they look new and well-maintained? If so, this is a good indication of a top-quality organization. Meet the pilot who will fly your group. Remember, if your group arrives at 6:00 A.M., this is the first person they will meet. Thus the pilot is in fact your respresentative, responsible for delivering your group to you in a happy mood. Make sure he is "a shirt and tie pilot" not a "cowboy boot and dungaree type." Explain to the pilot the purpose of the trip and that most of these people have not flown before. Brief him or her fully on the itinerary for the day, the names and positions of the passengers he or she will be carrying, and how to reach you if they are delayed or have a change in plans enroute. Have the pilot see that coffee and donuts are provided for the passengers. One of the keys to successful travel arrangements is to anticipate all possible problems and have alternate plans for each one. Make sure that they can contact your secretary from every stage of the journey.

Now that you have the travel planned and have set up alternatives to cover any disaster or problem, what can you do to make the trip as personal and pleasant as possible? You should be the person they call as they depart, telling you what time they expect to arrive so that you can assure them that you will meet them personally and that the trip is still on. There is no substitute for this personal contact.

If you can arrange your schedule so that you are in their area the night before, you can ride with them to the site. This would really make the trip personal and give you an hour or two to talk with them and to build a working relationship. Most of the conversation should be general, with only a brief review of the project to be visited. If you say too much about it in advance, you will have nothing to say when you are actually at the site. If you ride to the site with them you will probably accompany them also on the return journey. If you do, they will only talk about the good things while you are there. But if you are not with them, they may be critical. Keep the conversation positive until the group breaks up. It will probably be several days before they reconvene, and if they have not had the opportunity to reinforce and record any negative thoughts, they are likely to forget them and remember only the positive things.

Make sure you program a break in the first half of the journey. Pass out some material for their review while they are traveling. This should include a detailed schedule of where they are going and what they will see. The schedule should be an overview of the day's activities complete with a time schedule for each item. This will give them the total picture; knowing the time allotted for each part in advance will help keep them on schedule through the day. Include information on the background of the projects they will visit. This should include floor plans, elevations, costs and other information. Include photographs and any site plans available. The more they know about the project in advance, the more meaningful the visit will be. An additional piece of information that is always of interest is a listing of the people they will meet during the day. This list should include their position and relationship to the project and any other interesting facts. When they meet these people, conversation will then be easier.

The last thing to remember in travel is to acknowledge the time lag involved in even a short trip. Often we stand at the gate waiting for the group to arrive. As soon as they step off the plane, we start on a whirlwind tour that only ends when we put them back on the plane. If you watch the people closely, you will note it takes some time for the group to catch up with you in their thought processes. They are in a strange environment, have gotten up earlier than usual, and traveled for several hours. They need time to unwind and adjust to the new setting. Take a few minutes and go for a cup of coffee. This will not only give them time to relax and get ready for the business at hand, it will also give them the opportunity to meet the people from your office who will be with them for the day.

Selection of buildings to visit is important. If the prospective clients are expecting to build a school for about 400 students, do not show them a project for 2000 students. They will get lost in the complex and be completely overpowered by it. They will feel their project is too small and that you really are not interested in it.

I remember being invited to speak to the New Hampshire Vocational Education Directors on the subject of new trends in vocational education. I had researched the audience in advance and found that most of the directors would be needing small additions to their present facilities and that this was the main purpose of the meeting. My presentation was therefore built around a discussion of new trends in rural vocational facilities. The other panel speaker was an architect from a large metropolitan area. His entire presentation was about new facilities, the smallest of which was to cost many millions of dollars. It became obvious at once that he had missed the audience and was talking about an area they could not relate to. The questions at the end of his presentation were of a general nature, whereas the questions at the end of my presentation were specific, relating to their forthcoming projects.

Make sure the projects you show have some relationship to their projected buildings. Even if you believe the elementary school you built to be better than your high school, if they are building a high school, show them that as well.

The cost of the projects shown should be similar to their project. Take the costs of construction of the buildings you are showing them and update them to today's costs. Often owners will come to you with a very limited budget and will want an inexpensive building. They will generally make this known to you prior to the interview. When they visit your project, they will assume that all the details and finishes they see are the ones they will get in their buildings. So do not show them a Cadillac if you know they only want a Ford. When you deliver the Ford, they will be upset if it doesn't look and drive like the Cadillac you showed them. Stop and think about it. How many times has one of your clients said as he or she walked through the completed building, "It doesn't look like the building you took us to visit before we hired you!" If such a comment is not handled delicately and properly, you risk losing an owner and a major source of new work.

Be sure the buildings are ready for inspection. Do not make the mistake I made once when I called the owner of a building to tell him I wanted to bring a group of people to visit his building, but neglected to go to the building before the visit. The visit was disastrous. The building was dirty; it was vacation time, and all the furniture was piled in the corners of the rooms. As if this were not bad enough, the director of the vocational education department had just been fired by the board, causing the remaining staff to behave in a disgruntled and negative way in their conversations with us. As a result of this visit, we lost a major contract for a very large vocational educational center, all because we failed to inspect the building we selected for the visit.

Do not commit yourself to visit a specific project until you have personally visited and made certain it is the right one to show and that it is ready for inspection. Check to make certain that the building is clean inside and out, and will be in use during your visit. Do not visit a building that has unresolved problems. You might as well send the group home and save yourself time and money.

While you are checking out the building, talk to the staff. Make certain they are still happy with the building and will give you a strong recommendation when you bring your group through. If you detect an area of unrest and dissatisfaction among the staff, think twice before you bring a group through. If the staff is unhappy with either the building or the administration, this will be conveyed to the visitors and may cause a problem.

Do not let any prospective client visit one of your buildings in your absence. You need a successful, trouble-free project in order to get new work, along with a happy, satisfied client. If either of these are not available on any of your projects, do not show or use them as a reference. Start today to clean up all your existing problems and keep the owners happy. Then you can use any of your past projects for visits without fear of the owner giving you a bad recommendation.

During my travel and research for this book, I very seldom found owners who were happy with their last projects. The problems and complaints were generally simple irritations that could be corrected if the architect or engineer had given more attention to details. The problems were seldom costly to correct. When I did find owners who were happy with their building, I also found that they were giving all their new work to the office that had made them happy. The easiest way to get all the work you can handle is to make your present and past owners fully satisfied with your work. Not only will they give you all their additional work, but they will recommend you to their friends.

Once the problems are cleared up and the building is ready to be shown, how do you get ready for the visit? The visitors will be unfamiliar with the setting you are taking them into. They have probably already visited several buildings done by others thus adding to their confusion.

There are several ways to help your visitors from being lost in your building. As mentioned, copies of the floor plans should be given to them to review as they travel to the project. As soon as they arrive at the building, take them into a conference room and review this with them. This will be the time to hand out your material and discuss the details of the construction, such as costs, schedule, and other items. You also can introduce them to the key people who will be available at that building for questions and answers. But best of all, this is an excellent time to familiarize them with the building. Many architects or engineers will have a set of working drawings available and will use these to make visitors familiar with the project. Remember building and working drawings are new to these people and will probably confuse them.

The easiest solution I found was to make up a simple line drawing that could be reproduced on 8½ × 11 paper. Copies can be given to each visitor to carry with him as he tours the building. Have the local blueprinter blow the 8½ × 11 drawing up to poster size, and you will have a large display and talking board. You can color code the exhibit to help them understand what it is they are going to see. A word of caution: Do not give one of your draftsmen a set of working drawings and tell him or her to make up a floor plan. I have yet to see a building where the walls and rooms were built exactly as shown on the working drawings. It might be as simple as the relocation of a few doors or as major as a complete rearrangement of rooms. Take time to check the layout with what is actually in the building. If your plan does not agree with the building, it is another point against you in your struggle to establish your credibility.

Do not forget to include the name of the building on all the material you produce as well as the location of the building and the name of your firm. You never know where these handouts will end up, and it is good to have the project identified and to receive credit for it.

Resist the temptation to show too many buildings. A good rule is to visit one excellent building rather than try to rush them through two others. Give the group time to travel and eat. It takes longer to get a group of people started and to keep them on course, than it does to go through the building by yourself. It is important that the visit be smooth and unhurried. Keep it moving so that they feel they are accomplishing a great deal. You are familiar with the building and could go through it rapidly, but they are seeing it for the first time. Give them time to absorb what they are seeing and to talk to the people they meet. They will want to talk to teachers, custodians, and everyone else they meet. Give them time to do this.

Who should be present for the tour? You want to have enough people there to field any and all questions. Your visitors may be interested in broad-scope answers and also may be looking for detailed, indepth answers. For example, if they ask, "What is the program for this school?," it is not enough to answer, "a high school for 1200 students." They may want to see the actual education specifications and talk to the people who wrote them as well as the people using the space. Be prepared for this type of question.

If you personally were not involved in the project you are showing, arrange for the project manager from your firm to be there. This will make available to you the person who has all the answers. It will also give you an additional opportunity to show the depth and expertise of your firm. If you are not a partner in the firm, but will be running the project if your firm is selected, it is of paramount importance that the partner be present with you to back you up and signify that the firm will stand behind what you are doing and saying. A prospective client generally wants to see the partner during these preliminary meetings and visits. I know we lost several prospects because some in-house scheduling problems prevented the partner from attending these preliminary meetings. The owner told us later that this was why we were not selected. They felt that if the partner was not interested enough to follow up before the contract was signed, they certainly would never see him afterward. They were always willing to accept the manager as the person who would do the work and would be their prime contact to get the project completed. But they want, and rightly so, to see and get to know the person who will sign the contract and bind the firm to performance. Often during the course of the project, managers leave or do not perform and the owner wants to make sure the partner is a person who will continue the project as they wish.

Another key person to have on a building visit is the partner or staff member who was in charge during the design and construction of the building. This person will have important things to say about the programming and design of the building.

Another form of visit is to invite the prospective client to your office. This is especially effective if you are a large, full-service organization and want to impress him with your in-house ability to provide total services. It is to your advantage to get the client group into your office, where they are on your territory and you can set up displays and other exhibits to establish your area of expertise and past performance and show them the volume of work you have done. It is also a good time to get them involved in starting their project.

Travel arrangements and meals should be handled in the same way as for a site visit. Alert your staff to the visit. Tell them you are bringing a group through and they should be prepared to meet them. Make sure the schedule is not in conflict with any office functions that would upset the visit. This is a good time to have a general house cleaning. Get your staff to clean up and throw out all the things that have been accumulating, sometimes for years. It is impressive to walk into an office and see everything in its proper place. Cluttered work stations do not impress prospective clients, especially when you are trying to impress them with how efficient and well organized you are.

Select with care the office displays you wish to exhibit. Do not try to put up examples of all the projects you have done. This will give a cluttered impression. Rather, select a few outstanding examples and place them in strategic places so that they look like working displays rather than special exhibits. These displays should be as similar as possible to the kind of building the client group is interested in, and you can emphasize these similarities as you explain the display. This will probably be the first step the clients will take in trying to visualize their projects. This is good. If you can get them to dream a bit with you, they will often build their project with you. Try to have the best and most appropriate display set up in your conference room where the tour culminates. Seat the group around the table and have your secretary bring in coffee. This is often the time to close the deal. Have available some large-scale charts to show how your office works and how their project will be handled from programming through completion of construction. As you review this process with them, you can relate the conversation to the project on display. What you have done is to get them to start building in their minds and to commit their thoughts to you and your firm. If you do this correctly, in a low-key way, you will probably obtain their contract.

Brief your staff members on the nature of the visit, the type of building the group is expecting to build, and their general interests and background. This will allow your staff to prepare, intelligently, for the visit and to anticipate questions. When your guests are introduced, your staff members will know about the visit and will make them feel welcome. This emphasizes the personal approach and enables the prospective clients to visualize these staff members working on their project. Avoid the casual "Hello, how are you" kind of introduction.

Two other types of visits can be successful prospecting efforts. One is an open house at your office; the other is an open house at your recently completed project.

If planned properly, an open house can be very effective. You might time it to coincide with a state or regional school or health facility convention and invite a group of the participants to visit your office and meet your staff. If some of these people are current or past clients, this gives them a chance to renew acquaintance with your staff; and if they bring guests, it gives your office additional exposure. This is good public relations.

This is a social, not a selling situation, and you should avoid anything that might be construed as high-pressure sales. If there are several clients —either past or present—in the group, exhibit their projects. An owner's project is like his or her name. He or she likes to see it displayed. When he or she notices it, he or she will show it to their friends. This is excellent public relations for your firm.

When you open a new office in another town or remodel or enlarge your present office, have an open house. This is time for a party and should be so treated. Invitations to all past and present clients will give them an opportunity to look at your new facilities and help celebrate your growth.

Arrange the office for the visit by displaying examples of a cross section of your work. Choose colorful displays. Have only your key staff and secretaries there. Invite the wives of your staff as they will add greatly to the festivities. This will probably call for an open bar and some hors d'oeuvres.

If your open house is for the opening of a new office in another town or state, it is important that the local people get to know you and your staff. Go over your invitation list carefully, and be sure all the local dignitaries are invited. You might also consider a separate luncheon before the open house for them to meet the key people you have brought in from the home office. This will give them an informal advance meeting of your team and will insure that they come to the open house. When they arrive, they will already know you and some of your staff.

My favorite kind of open house is one that takes place after the grand opening of one of our projects. This is an easy one to justify and one that excites people into coming. This is not the same open house as the one owners usually hold to show the local people the completion of their buildings, but is set up by the architect after the building is in operation and all the problems have been resolved. This is the time for having as guests local and state officials as well as clients and prospective clients.

The owner should give an official greeting and if it is a state project, a state official might be invited to extend his or her greetings. Give your guests the background to the project and especially bring to their attention the new ideas and techniques that are present in this building to make them aware of its unique qualities. Have available for them details of the costs and schedules if appropriate.

Transportation

Transportation can be a favorable prospecting tool or a disaster. Your number one means of transportation is your private car. Your car must be in top shape at all times, because it can mean your survival. Most of my meetings lasted until the late hours. If your car will not start at midnight, it is time-consuming and disgusting to be in the parking lot by yourself, wondering how you are going to get started because a service station is hard to find at that time of night. I was usually about a two-hour drive from home. The drive was generally along deserted New England roads in the winter time. A breakdown could be a mighty cold experience. For this reason, I carried a sleeping bag, heavy boots, extra sweaters, gloves and hat in my trunk to keep me warm if I got stranded. I also carried shovels and flares. I kept my car in good condition and, fortunately, never became stranded or required this emergency equipment.

People you visit may judge you by the type of car you drive. If you drive an expensive car they may be impressed, but they may also think you must have made a lot of money from your past clients and wonder how much you plan to make from them! If you drive a medium-price, economy car, they may look on you as being in the same income bracket as themselves. No matter what you drive, it is important to keep the inside clean and dusted so that it is inviting. Many filling stations now have drive-through car washes you can use when you fill your tank. Each time you stop for gas, take five minutes or so and get your car washed.

We have talked about the use of the car for traveling distances of up to 150 or 200 miles. How do you handle distances between 200 and 500 miles? If you use your car for a prospecting visit to a distant area, you will have to allow four to six hours driving time before you can make your first call. To spend a full day prospecting, you will have to leave the day before and return the day after. As an alternative you might consider the private plane or air charter service. People's first reaction to this idea is that they cannot afford to charter an airplane. But if you take into account the cost of a 500-mile drive, plus the cost of hotel rooms, it is often cheaper to fly.

If you do a large amount of traveling, you will find as I did, that a private plane has many advantages as a prospecting tool. A plane allows you to cover a larger territory more rapidly and with greater efficiency. Until my office opened a branch in Bangor, Maine, I worked all the Maine projects and new business development from Burlington, Vermont. Driving to northern Maine would have taken ten to twelve hours. Flying there on commercial airlines required an hour flight to Boston, an hour layover, another hour flight to Bangor, renting a car and a two-hour drive north. With a private plane, I could leave my house at 5:30 A.M., be at the airport ready to take off at 6:00 A.M., and in my client's office or starting to prospect by 8:00 A.M. I could work a full day and still be home for supper. This had an interesting effect on the owner, who saw my availability on the job as a counterbalance to the trend to hire a local architect.

The airplane is not only an efficient means of transportation, but also a prospecting tool. Just as the type of car you drive is an identification of you, so also is the airplane. Combine two or three stops on each trip and you subsequently reduce the total trip's cost. Be alert to the travel needs of prospects and clients. If you have a meeting with state officials and you both have to drive, rent a plane and pick them up. They will appreciate it and will be your captive audience. People do not often fly in small planes and it is often a pleasure for them to do so. Use the plane as if it were your car and involve the prpospects in flying as much as possible, but do not bore or try to impress them.

Commercial airlines are the only means of travel for distances greater than 500 miles, but to avoid the possible trauma of delays it is wise to go the night before and avoid the worry of having someone meet you who may be kept waiting at the airport. When you have checked in at your hotel, call the person you are visiting to announce your arrival and start bright and early the next day.

Whatever method of transportation, allow enough time to arrive on time, and if possible be early enough to stop just before you get to your destination to have a cup of coffee and unwind from the trip. This needs to be stressed, as lateness is irritating and if you become known as someone who is always on time, you will do much to set yourself above the competition.

If you travel very much, you will undoubtedly sometimes encounter problems, such as being delayed by weather or mechanical malfunctions. Always anticipate the possibility of such delays by setting up a check-in system: when you leave one point, call your secretary and tell him or her where you are and what is your next stop. Your secretary can then call your next contacts and advise them of your expected arrival time. If you are running late, he or she can call your other appointments and advise them of your revised schedule. People do not like you to be late, but they appreciate being informed in advance. They can then go on with other business until you arrive. If someone wants to locate you, tell them they may call your secretary. They should also call your secretary if they want to change any plans. Owners or prospective clients will appreciate the respect you show them by this thoughtful communication. When you are talking to your clients about meeting them, go over your proposed travel plans, including mode of travel, with them, letting them know where you will be arriving from, time of arrival, and where you are going after the meeting. They may recommend a place to stay and dine, but you should always make your own arrangements.

YOUR SUPPORT FORCE

Your Employees

The most important attributes you can have in prospecting are your employees, a fact too few professionals recognize. Your employees must be made to feel important and encouraged to become strong supporters of your firm. Your office must be free of undercurrents that cause discontent, lower morale and undermine the firm. Use your employees. At sometime in everyone's daily life, the individual hears about some new project which is going to be built or someone who needs an architect or engineer. Every employee must be alerted to listen for this kind of information and pass it on to you. If the employees are strong supporters of you and your firm, they are going to talk with pride to their friends and relatives. Their friends will then be aware of you and when asked sometimes if they know a good firm, your name will come to mind.

Your Secretary

Look at your key employees and see how they can help you to prospect. The first and most important person to you and your work is your secretary. A secretary is the vital link between your office and your clients and prospects. When they call, the secretary is the one they will talk to first; if you are out, he or she is the only one they will talk to. It is important that this person become familiar with all your clients and get to know them personally. I used to make a point of taking my secretary to visit all the projects and meet all of the clients. When they called, she was not just a voice on the phone but someone they knew. The secretary must recognize each voice and make callers feel welcome and glad they called. If you are out, the secretary must sound confident and be able to persuade callers to leave messages. Callers must feel they can trust this person to deliver messages accurately. Your secretary must know the status of all your projects and also of all your prospects. Keep him or her up-to-date on progress. If new prospects call and identify themselves he or she will recognize their names, and by knowing their relationship to your prospecting work, the secretary can make them feel important. Help your secretary by keeping him or her fully informed about where you are and what you are doing. This allows the secretary to answer intelligently when someone wants to reach you on an important matter.

Your secretary is the firm. He or she may be the only contact people on the outside have with your firm. The way he or she answers the phone and treats people who call is of vital importance to you and your firm. You cannot afford to have a secretary who is not interested in making your firm the best there is.

Field Inspection Crew

Your field inspection crew is another often untapped resource for prospecting. These are people who are constantly in the field and have the opportunity to meet not only the owners, but also the general public. They should be trained in the art of public relations so they will know what is expected of them when they are out. They must conduct themselves in a professional manner. They must be strong supporters and representatives of your firm. During their visits to the job, they will meet many people in the construction business. These are people who are working on other projects. Your men should be alerted to listen to the conversation when it shifts to discussion of other jobs and also learn to ask leading questions about other jobs in which you are not involved. This information should then be fed back to the manager to help give him or her a picture of what is happening in the area. When fitted together with other information, these reports might result in tracking down a definite lead or the establishment of a trend in the local market. It is vital to your success that you know what is happening around you, who is doing what, whose projects are in trouble and why, who is successful and who might be considering new buildings in addition to the work presently being carried out.

Your field crew can also instill confidence in your firm on the part of the owner by demonstrating that you are giving him or her full service. Many field inspectors arrive in town, go directly to the job site, inspect the building and then leave for the next project. This is a one-sided visit, but with a little effort it can become a valuable prospecting trip. When the field inspector arrives in town, he or she should be available to the owner. He or she should answer any questions the owner has and make notes of information to bring back to you. The field inspector is an important public relations representative of your firm. Existing clients are more likely to continue to give you work and references if they feel that you and the members of your firm are readily accessible to them.

Train your people to stop and see the owners as soon as they arrive in town. They should explain their visit to them and tell them what they will be looking for in the inspection. This will show the owners that they are fully briefed on the status of their jobs. This is important. They should if possible make an appointment to meet the owners at the job to walk them through it and bring them up-to-date on the inspection. If during the visit the owners ask questions the inspector cannot answer, he or she should call the project manager at once to get the answers. Owners will be watching the inspector closely because he or she will appear to be a new team member, one owners want to be sure they can trust.

CHAPTER FOUR

MAKING TIME FOR PROSPECTING

The human machine is one that will not start itself. You must start it. Left to our own devices, it is too easy to sit around feeling sorry for ourselves and finding all kinds of reasons why we cannot begin a project *today*. Architects and engineers typically look at a job and immediately try to determine how long it will take to complete. They then determine the due date, subtract from the due date the projected amount of time to complete the work, and that is when they start working. They always seem to be racing the clock to meet a deadline. This is a bad habit to get into and one that should be broken. In prospecting and creative management it is better to start early and do more than just the minimum required.

ORGANIZING YOUR WORK

Reinforce Your Current Projects

The first priority in new business development is to reinforce your current project so that you will have time to devote to prospecting. Resolve job-related problems and set up your projects so they can run without you for two days each week. Never think of yourself as too busy to look for new work, but set up a strict schedule that gives you two days a week to devote to prospecting and then begin to make new business appointments.

The following is an outline to help you define what must be done to clean up your current projects and give you some ideas on how to go about it. If your projects have continuing problems, the clients will be unhappy, and an unhappy client will not get additional work for you.

1. CLEAN UP YOUR PAPER WORK. We are all faced with an in-basket piled high with mail. If you are like everyone else, there are many items to be handled, such as change orders, letters of approval and minutes of meetings that must be done. Time must be taken to seclude yourself in a quiet office, perhaps over the weekend, and start at the bottom of the pile and work until you reach the top.

2. CLEAN UP JOB PROBLEMS. Every project has problems of which we are all aware. Most of us look at the problems everyday, but tend to postpone action. Some problems may be embarrassing as they involve difficulties that we, ourselves, have created. But problems like these do not go away. In fact, the longer you wait, the tougher they are to solve; so you might as well attack them at once. Start by developing a list of all the jobs and all the problems you have on each job. Then develop the procedures you can use to solve each problem. Once each step is defined, assign a time by which it must be completed. Assign your staff members to start correcting all those problems that you do not have to handle yourself. In many cases, a new face with direct authority to handle a problem will accomplish more than you can yourself. If you are a good manager, you should be able to delegate work without losing control. Do not keep all the work to yourself. Insist that your employees do what you hired them to do.

3. REVIEW ALL CORRESPONDENCE FILES. This is an experience that few architects ever undertake. Go back to the inception of each contract and review everything that is in the folder. Make notes. If your files are typical, you will find many interesting things. First, you will probably find a number of letters you wrote to other people requesting them to do things. Some will not have been answered; others will show requests that have not been complied with. Make a note of all these items; if the requests are still valid, make a note to follow up. Next, you will discover letters and requests from other people, including contractors, which you probably have not answered or complied with their requests.

   Next, look for trends in the correspondence. On a typical project, who is writing most of the letters, you, the contractor, or the owner? If you are writing the most, is it because the others are not doing their job, or are you trying to build up a case to protect yourself from either the owner or the contractor? If the contractors are writing most of the letters, what are they trying to tell you? For example, contractors may write to ask you to do something, such as complete and return shop drawings. Then they will send follow-up letters asking why the shop drawings are not returned. These letters may be damaging to your position, if a project goes to arbitration or court and the contractor claims you held up the project. Be alert that contractors in their letter writing might be building a case against you or the owner for claims for extras to be filed at the end of the job or to counter charges that the owner often files against the contractor for nonperformance.

Third, watch for letters from owners requesting performance of tasks you have not done. Be alert to the possibility that the owner might also be trying to prove a case of nonperformance against your firm. In all these cases, a single letter across your desk generally does not excite you or alert you to problems. But when you look back in your file and find that you have received a damaging letter each week for the past two months, you are heading for trouble.

In summary, note first requests you have made of others, but that they have not done. Second, note requests that others have made but that you have not done. Third, note any trends that indicate that a case might be made against you. The problems must be solved at once.

4. DRAW UP A MASTER LIST FOR ACTION. Now take all the lists you have accumulated and combine them into one master list. This list should include all the paperwork that is not completed and all the job problems that are outstanding and must be cleared up. Include all the uncompleted items revealed from your review of the correspondence file. Now arrange everything in order of priority and set deadlines for each action to be taken. If some of them can be assigned to your staff, so much the better. Get your desk cleared and your house in order. Now place a copy of this list on your desk and give one to your secretary. Each week, he or she can call or check with the people to whom you have assigned work. As a result of these calls, the secretary will then present you with a status report on what has been done and what is still pending. Use your staff as much as possible to help you deal with this backlog of work and get out into the field where you can be

most effective. In many cases your staff who are not involved in all the emotional overtones of procrastination and guilt will get these jobs done more quickly and efficiently than you would yourself.

More Efficient Use of Your Time

Once your desk is cleared off, perhaps for the first time in years, it is time to take the next step in learning how to make efficient use of your time. If you are getting your job-related work done and also spending time out in the field developing new business you are already getting more productivity out of each day.

One of the methods a project manager uses to achieve more productivity during his typical day is to get out of the office into the field and become less available. If the manager stays in the office too much, he tends to get too involved in the details of running the project rather than delegating these to his staff. If you are constantly at your desk, your employees will not make any decisions without stopping to consult you and then coming back to advise you about the results. While it is desirable for them to keep you involved, it is not if the conversations take up too much of your time. If you are to be out of your office for the day and you have a secretary and a staff of people working for you, you should advise them the day before that you will be out and that you will call the office at, for example, 8:00 A.M. and 4:00 P.M., when they should have all questions and messages ready for you. This will force your staff to plan ahead to keep you advised and to obtain the right level of involvement from you. The 8:00 A.M. call will insure that they came to work on time and it will probably not last over twenty minutes. You have thus freed the hours between 8:20 A.M. and 4:00 P.M. to do whatever you think most important in new business development. If you were in the office, the phone would be ringing, people would be coming into your office, and you would never get any quiet time to concentrate on new work.

Get out of the office as much as possible and organize your team by phone. If you are an efficient long-distance manager, you will find that you will get excellent work from your team. By getting out in the field, you will be able to spend more time with the owners on the job site and prospecting for new clients. I strongly believe this is where the manager should be. The more owners see you, the harder they will think you are working for them. Owners typically have the misconception that the only time you are working on their project is when you are in their office or on the job site. They all feel that when you leave their project, you are going on to work on someone else's building. Let owners see you as much as possible. It will help your image.

Make meetings and visits productive but short as possible. We could probably reduce by half the length of each meeting and get more accomplished. Meetings may be necessary, but a great deal of time is often wasted on unnecessary detail. Because of a lack of leadership and that often several people are late, most meetings start late. The chairperson may have neglected to make up an agenda and consequently has to spend valuable time telling those present why they are there. More time is wasted if the chairperson is weak and does not control the schedule. If the first item on the agenda involved only two of the people in the group, the rest have to sit and cool their heels waiting for their subjects to come up; so the meeting is probably only 80% efficient in the use of everyone's time. In typical meetings of one hour or longer, after the first forty-five minutes, the discussion often becomes ineffective. A worthwhile meeting can only be achieved by planning ahead. Advise the participants what topics are to be covered, prepare your remarks, and then run an efficient and often, short meeting.

Plan your time every day. Often, one tends to arrive in the office in the morning, sit down at the desk and start doing whatever happens to be in the work basket. Once that is finished, the rest of the day is spent doing whatever happens to come up. By the end of the day you will be tired from responding at full pressure to each demand that was made of you, and you may feel that you have not really accomplished anything. You must set up a work schedule for yourself everyday of the week. Learn to control your time and to complete a project when you start it rather than acting as the office juggler. We often do not realize this is happening to us until it is too late. Listen to your staff. They are usually aware of it before you are. One of the occupational hazards of successful managers is that the more successful they become, the more work they get, and the more work they get, the more they have to juggle. Some managers feel that when they commit their firm to a job, they must do all the work themselves. But it is impossible to do everything yourself, and owners will be upset by the delays this may cause.

When I was training project managers to work under my direction, I would attempt to have them plan their time and travel so that they were as efficient as possible at all times. If a job problem or a travel detour occurred, they would have an alternate plan and could keep a smooth work program operating. They understood this and became efficient, but occasionally, in the event of a major disruption, I would have to call them wherever they were and get them to change their plans to stop what they were doing and go somewhere else and do some completely unrelated task. It was then that difficulties arose. They found it hard to understand a change of plan, even though at one level they all knew that the only constant in life is continuous change. Change must be accepted as an opportunity to take off in a new direction and must be welcomed rather than resisted. But you should become concerned if you are constantly changing your plans and when you look behind, all you see is a group of frustrated people standing there waiting for you to return. If this is the case, you are not reacting to genuine changes, but have become the juggler again with too many balls in the air at once.

Suppose you have scheduled a meeting with one of your clients some distance away and have stayed up late at night getting the project completed and making your travel plans, but on the day of the trip, just as you are ready to leave, the phone rings, and the owner tells you the meeting is canceled. How do you react? The first inclination is to tell the owner how much he or she has put you out, but this temptation must be resisted, for such will only embarrass the client and do you no good.

Instead, try to reschedule the meeting. And as soon as you have hung up the phone, get immediately to work on your next project. Never sit around and complain about the wasted effort or the crash program you and your staff went through. Rather look on the canceled trip as providing you with a large block of unscheduled time you can now devote to other pressing work. Think positively and flexibly about such situations and do not let them disrupt your working rhythm.

Learn to make every minute and every effort count. Is your office an efficient place to work, or do you spend a lot of time walking around, looking for things? This may be more time-consuming than you realize. Looking up telephone numbers is another time-consuming detail you should avoid by an up-to-date file next to your phone.

Avoid grouping too many desks in one area, especially if people spend most of their time on the telephone. Separate the desks and give them a quiet place to work. You will be surprised how this improves morale and productivity. Be alert to a confused working environment as a major cause of lost productivity and do something to correct it.

One of the next things to remember in this effort to become more efficient is to remind yourself and your staff that the working day for everyone is 8:00 A.M. to 5:00 P.M., or whatever, with a fixed lunch hour.

Watching people in a motel dining room is an instructive lesson in effective use of time. At 6:30 A.M., a group of salespersons arrive for breakfast, wide awake and alert. They look efficient and sharp. By 7:15, they are out of the dining room and on their way to their first appointment. Dining is slow until 7:45, when the next group arrives. These people stand at the door, wondering what to do. They eat slowly, and drink several cups of coffee. About 8:30, they get up and go back to their rooms and get ready to go out for a 9:00 o'clock appointment. Around 8:30, the third group arrives. These are the ones who were out last night on the town. It was hard for them to get up and get started. They have a couple of cups of coffee before they can even order breakfast. It is evident that this group will not be out of the motel and at their first call before 10:00 or even 11:00.

Watch in the evening and the whole thing happens in reverse. The group that left around nine for the ten o'clock apointment checks back in around three or three-thirty, because no one wants to be bothered after three in the afternoon. The people in this group are lucky if they got in three or four hours of productive work in the day. Around four or four-thirty, the second group arrives. Even though they agree with the first group, they did make their last call at three-thirty and probably did get four to five hours of work in during the day. Then around six o'clock the largest group arrives. They come in with full briefcases. They did not leave their last appointment until after five o'clock. They will probably work for an hour or two in their rooms organizing their material from the day's calls and planning the details of tomorrow's work. Which group would you want to work for you, and which group do you belong to?

Success demands dedication. You should make your first business contact at 8:00 A.M., or when your office opens—whether it is a meeting in your office or your client's or your first prospecting call. If you work to 12:00 noon, start again no later than 1:00 P.M. and do not leave until 5:00 P.M. This work schedule alone is likely to put you ahead of the majority of other firms.

As you plan your week activities, group them to allow you to reduce your travel time and make one trip instead of two. It does not make sense to drive seventy-five miles to a town to visit one of your projects and then return home, driving that same seventy-five miles, without accomplishing some work along the way. The time and cost of driving 150 miles is enormous and should be profitable. Rearrange your schedule to cover other work on the trip and use the time saved to develop new work.

As you go out on the road, you will find that you are spending more time traveling and eating. This time is partly wasted if you do it on your own. Try to arrange your schedule so someone else can travel with you. For example, if someone else in your office is heading to the same city or job, invite that person to ride with you. It will give you a chance to discuss ideas with the person and get to know him or her better. If you and a client are going to the same meeting, pick him or her up on the way. It will give you a chance to plan the meeting and get to know each other. The same is true when you dine out. Try to meet someone for breakfast, lunch, or dinner. This not only can turn a lonely experience into a pleasant one, but it is enjoyable for the person you are entertaining. Once again, you have gained some more time in your day.

Involve owners daily in your decision-making process. This will save time when it comes to problem solving. If owners have been involved in the decision-making process from the start and are aware of all the steps along the way, they will be conscious of the problems and perhaps become deeply involved in them. Once they understand the background of the situation, they can help come up with a solution. They will take it to the board and get approval for you more quickly because they have command of all the details and are on your side.

A manager has two functions. The first is to keep information flowing into the office so that the staff continues to be busy and does not have to sit around waiting for directions. The manager's second function is to solve problems, through creative thinking, and to plan the overall operation. While the first responsibility has to be handled in the office, much of the creative thinking must be done outside. While driving your car to the next meeting, try to concentrate on solving a single issue. Have a small, pocket tape recorder with you so that as you think of solutions, you can put them on tape. To think and plan, you must get yourself out of the mainstream of the activities and into a position where you can give exclusive attention to only one problem for an extended and uninterrupted period of time.

Another time-saving habit is to keep accurate, written records; these depend on your ability to take complete notes. Consistent record keeping along with complete and up-to-date files are essential. If you have to spend a great deal of time each day looking for misfiled letters, this is time wasted that could be better spent prospecting. If you reach an agreement with someone on the solution to a difficult problem, but fail to record the agreement, it will soon be forgotten and the whole process will have to be started again. Once again this is time wasted that could be better spent in seeking new business.

Always take the initiative in correspondence. If a decision is reached at a meeting, as soon as you get back to your office, send out a confirming letter to the people involved. Do it the same day if possible. The reason for this is simple. Most people dislike writing and sending out letters. If you wait for the other person to write the letter, several days may elapse before he does it. After several days, people tend to forget exactly what they said, and more time is wasted trying to clear up confusion over the matter. If you let someone else write the letter first, you are subject to his interpretation of the matter, which may not agree with your own understanding of what was decided.

Much time is wasted if we fail to develop and use standard formats for common purpose letters and transmittals. In some offices, all transmittals for processing and shipping shop drawings are still being individually written, when all that is needed is a form on which one or two boxes have to be checked. There are also many typical letters that we recompose every time we need them. Why not develop, from your files, a notebook of typical letters that can be left with your secretary. In this way, when a standard letter arrives which has to be processed through your office, your secretary can process the paperwork and place it on your desk with the incoming mail so that as you read, you can review the response he or she has prepared. You can either sign it or alter the response. If you have set this up and properly trained your secretary, you should be able to review and sign 80% of everything that he or she does; 10% might have to be corrected and 10%, probably rejected. You will have greatly speeded up the flow of paperwork across your desk.

All salespersons who walk into your office will try to convince you that their equipment can pay for itself and will therefore cost you no money. Unless you can actually reduce your payroll by the amount of the rental and other related expenses, beware! You might find yourself with the same payroll and no increased production, but an increased monthly overhead because of equipment cost and rental. But good inexpensive equipment for your personal use can help to free your time for prospecting. The following is basic equipment that every manager should have.

1. DICTATION EQUIPMENT. You need a simple, lightweight pocket-size recorder, one that takes a fifteen-minute tape and can be transcribed by your secretary using a simple mechanism. On the way home after a meeting, you can dictate the minutes into the recorder, as well as all correspondence resulting from the meeting and notes on things to follow up.

2. TELEPHONE. If used efficiently, this is a most important piece of equipment. How often in today's business world do people complain that the phone is taking up too much of their time? Yet they want the phone to ring, just as they want to call other people to make certain they are following instructions. You must use the phone to stay in touch with, and on top of, all situations in which you are involved.

   Examine your phone bill. A large monthly bill is a good business investment only if most of your phone calls are three minutes or less. If they are over three minutes, you are probably talking too much and not sticking to the facts.

   Organize your phone calls before you pick up the receiver. Know whom you want to talk with and what it is you want to find out. Then place your call. As soon as you have the other person on the phone, identify yourself, the name of your firm, and the project you want to talk about. With these three statements, the person you are calling is now ready to receive your message. His or her answer may then be short and complete as well as to the point. If the person starts to digress bring him or her back to the subject. Use the phone to lead the activities and make people do what they said they would, when they said they would do it.

3. SPEED LETTERS. These can be assets in assisting in the recording of information and distributing instructions without the time-consuming process of typing and formal distribution. I strongly support this type

of paperwork for in-house communication, but do not send them to an owner.

4. POCKET CALCULATORS. All architects, engineers and businessmen need these when work demands proficiency with numbers. Owners want correct answers while they are talking to you. If an owner asks you "how much it will cost to build a project," you must be able to give him or her a reasonably accurate estimate. The pocket calculator is helpful when the problems involve such processes as multiplying the square footage of the building by the cost per square foot and then adding a percentage for inflation. The inability to make such calculations efficiently in the presence of the potential owner is not only embarrassing but also time-consuming. With a pocket calculator in your briefcase, you will have the correct answer in a matter of a few seconds and can quickly check to make certain you did not make an error.

5. PHOTOCOPY MACHINES. To gain additional time in your day, these are vital tools. Whenever you are about to write a long letter or report or to assemble some facts in a chart, stop and think if you have written a similar document before, which you can photocopy and mark up. The photocopy method combined with the cut and paste approach to assembling and redrafting will save time.

Support Force

As you move into management positions, one of the biggest problems you will encounter is that of letting other people handle what you did in the past, to free you for your new responsibilities. In architecture and engineering, this is as difficult as in any other business or profession. In fact, it is often worse. Since architecture and engineering are very personal, the handling and carrying out of details is not easily delegated to others. The pride of authorship is strong among architects and engineers, and they are reluctant to give up to others the right to make decisions on their jobs. But if your firm is to grow and you are to become a real and productive manager, this is what you must do.

If you think that the only time things get done is when you do them and that the rest of the time people are sitting around waiting for a decision from you, this is a sure sign that you are refusing to let people help you. If the people you have working under you cannot be trusted to help you, you should not have them on your staff. But the chances are that they could be productive, given a chance. You must use them or let them go. First, sit down with them and find out why they do not seem to come up with the same solutions as you do. Second, accept the fact there is always more than one way to get a job done. Examine their solution and see if the final result is as good as the one you would have produced yourself. Forget that their problem-solving approach may be different and concentrate on the quality of their solution. Open your lines of communication with the people working for you. Give them clear, concise directions. Let them know what you want them to do, when and how you want them to do it, and all other important facts. You must give up some of your work to your subordinates, if you are to find ten to twelve hours' free time each week for the development of new prospects.

If the people in your office are not immediately capable of assuming job-handling responsibility, start by letting them become your detail and leg people. If they do all the detailed work, gathering information and assembling documentation of facts and information for your use, they will soon grow in experience and proficiency. The Environmental Agency in Vermont has an "Act 250," the first of its kind in the United States. To start construction, you have to apply for approval under this act, and this requires the assembling of a large amount of project documentation, which has to be presented at several hearings and public meetings. To prepare for each hearing would have taken several days of my time, so I assigned one of my employees who was extremely careful about details to gather all the information and prepare all the reports and applications and become familiar with the facts. At the hearings and public meetings, my assistant sat behind me and passed additional information to me as I needed it. This was an unbeatable combination, and both the owners and the state agencies congratulated us on our thorough presentations and attention to details.

Another way in which your staff can help you is by making certain you do what you have promised. If in the notes of a meeting you state that you will present certain material at the following one, your secretary should note this on his or her schedule and help you make sure that it is done. This kind of follow-up not only reduces your work load, but relieves you of the burden of trying to stay on top of everything other people are preparing for you. Unless you are really a one-person firm where you do everything yourself, including the typing and cleaning, you have a staff whose job is to support you and make you and the firm look good.

Develop checks and balances to insure that things get done when and how you want them, and remember that this is a two-way operation. You are no longer working by yourself. You must build confidence in your team and trust in each other, but be careful not to give away too much of the action and control until you are ready to accept the results of someone else's decision making. If they are making the decisions, but you are responsible, there may be problems when an error in judgment is made by one of your staff. If a backup is brought in and assumes too much responsibility, owners will become confused about who is in charge and will not know to whom they should be talking.

Do not ever neglect owners while you are in the process of freeing up your time to allow yourself to develop new business. One of the most common complaints owners have about architects and engineers is that they are too ready to turn over their projects to a less-qualified staff, while they are out looking for more work. Do not lose contact with your clients, or your present owners will quickly become dissatisfied and you will gain the reputation of a fast salesperson, who relies on his or her staff to do the work. This is a bad image that is almost impossible to erase.

Proper use of support personnel will allow architects/managers to emerge from their load of detailed work and get out into the field. Because they will have more time and a fresh approach to the problems, their supporting staff can look at problems objectively and bring the necessary recommendations to the manager, who can then make problem-solving decisions. When you delegate work, be sure owners understand what you are doing and reassure them that you will still be actively involved in and responsible for their projects. Without this forewarning, they may panic when they see someone handling your work and think you have left the job. Owners fear losing the key person who started the project; so you must preserve continuity at all costs.

Organize Your Activities

The best way to organize your personal time is to establish the goals you wish to accomplish for each month, week, and day. Set out in broad terms the routine things that must be accomplished; then establish new items, such as the number of prospecting calls or different places you want to visit as potential new business areas. These goals and existing job assignments should then be scheduled for specific days in a specific week. You may feel that this kind of exact planning is not possible in the variable working life of an architect or engineer, but if you do not become goal oriented, it is all too easy at the end of the month to find that you have accomplished nothing.

It is a good idea to take an hour or two at the end of each week to get out of the office and quietly review your progress and accomplishments during the preceding period. Did you do everything you had on your list, or did unexpected events and pressures cause changes or postponements? Taking this review into account, you should look ahead at the coming week and reschedule your activities to take account of the items that have been overlooked. If you program your time hourly, not just daily, you will find yourself in control of your own schedule instead of, as too often happens, letting it run you.

CHAPTER FIVE

PROSPECTING SUPPORT FROM YOUR FIRM

Your firm and those who work there are a vital key to your success as a manager and a prospector. We talked earlier of how they can relieve you of detailed project management tasks, but they can also give you valuable support in the process of converting a prospect to a contract.

WHAT IS AVAILABLE

Research

You may have a lead for a master plan for the development of a park area. You may never have done this kind of work before, but on your staff is someone who has. Before making contact with the owner, call on your staff member to give you background on projects her or she has done, costs and cost updates, new trends and directions as well as what the competition is doing so that when you talk to the owner, you can appear to have a good grasp of this special field. You may display a strong working knowledge of the subject, but tell the owner that your staff member is considered an "expert." If the owner is interested, he or she will want to meet this member of your staff to discuss specifics of the project.

It is just about impossible for you to keep track of what the competition is doing in all the fields in which you are interested. However, this is another area of research for your staff to undertake. Give them directions to start researching the competition and to develop indepth reports about firms and projects. They should know who are the leading firms and individuals in your area in the different building types, what firms are presently doing the work, how their costs compare to yours, what time schedules they keep, their reputation and what new ideas and techniques they use. This type of feedback, when given to you prior to a meeting with a client on a specific type of prospect, can be invaluable in showing the client you are on top of what is happening in the building industry. One word of caution: when you discuss what you are doing in your firm and what the competition is doing, never discuss the competition in unfavorable terms. If you are doing something the other firm is not, do not compare yourself with them. Simply tell the owner that this is a service your firm provides. He or she will find out soon enough that the other firm does not offer this service. Never degrade your peers, it will reflect badly on you.

Technical Skills and Knowledge

It is impossible for you to be an expert in all areas of knowledge; so you will need to surround yourself with experts who can give you the information you require to be an effective leader of the team and demonstrate working knowledge of all areas as needed.

A typical owner's question is, "How much will this project cost?" One response—rather characteristic of architects—is that there is no answer until the owner has a program or a budget. But this does not satisfy the owner, who wants to know how much he or she is going to have to spend and what his or her money will buy before signing a contract. Your staff must be able to give you budget figures with built-in contingencies to cover the unknown. Even with the contingencies, the budget must be clear, understandable, accurate, and realistic. If you give the owner a budget and then he or she hires you, you will be expected to perform within these limits. Know the rules of thumb and how they relate to costs so that you can give the owner accurate estimates. If you do not, he or she will go to a contractor who will supply him with a figure.

Schedules are critical to an owner. When an owner asks whether occupancy, additions, or renovation can be completed by a certain date, give him an answer you can live up to, if you get the contract. Know how long it is going to take you to program and design the building as well as to produce and bid the documents. With the current emphasis on construction management and fast-track scheduling, you must know all the details of accelerating schedules and their implications for this particular prospective job.

The days are gone when owners would sign a contract and then accept the budget and schedule given by the architect later in the project. Owners are demanding that architects commit themselves to performance in terms of costs and schedules before the contract is awarded, and these costs and schedules may then be bound into the contract and cannot easily be changed later. Your staff is an invaluable aid to you in researching and double-checking this information.

Details of construction are a subject of concern to owners during the prospecting period. Owners are still saying that architects cannot build leak-proof roofs!

Study roofing details as well as other building details so that when an owner asks how something is built, you can take out a piece of paper and draw a section that will show him what you are talking about. Make sure that the details you draw are correct because if one of the owners is an expert in the area, her or she may ask questions just to test you.

Mechanical, electrical, and plumbing systems are areas that present massive problems and expense in buildings, but are often passed over quickly during the interview stages. Owners now into their second building project, will be on their guard and want specific answers in these areas. If the prospective job has an unusual mechanical requirement, involve your staff and take them to meet the potential owner either during the prospecting phase or at the interview. You are not, of course, to give the owner a solution prior to being hired, but you and your staff must know enough about the problems and concerns of the owners in this regard so that you can give a presentation of the steps your firm will use to solve these problems and the checks and balances you will institute to assure proper performance. Many owners feel that engineers and architects do not talk to each other or to them about mechanical problems, so you must show that you strongly support each other and have good communication.

Site engineering is in everyone's mind. Thus during the interview and selection process, you will be asked such questions as: How will the building be situated on the site to best exploit solar energy and conservation potential? How do you design and get approval for subsurface sanitary systems? How much acreage is required for a high school athletic program?

Another area of concern is subsurface investigation to determine the type of foundations the building will require as well as the cost of the foundations. Another subject often aired in early discussions is the well system for a large project. What happens if there is not sufficient water on the site? What can be done? These are just two of the many questions you must be prepared to answer for each prospect.

Owners are not interested in generalized answers, which may appear to be stock answers you have been trained to give in response to certain questions. They want facts that relate to their own sites. Involve your staff. Brief them on the background of this particular project so that your answers and discussion can relate to the facts of the project, not to general theory. If you are to stand out above the competition you must introduce your potential clients to a well-qualified and well-briefed staff to show owners that their projects will receive highly skilled professional service at all levels. At this stage, it is a good idea to give a detailed description of how your firm goes about obtaining all the environmental and other approvals necessary to construct and occupy a building in today's context of complex government regulations and controls. Special services and requirements are sometimes the key to the success of special projects.

The same kind of approach should be used for any special services that might be required. Some of the most common ones are food service, acoustics, hardware, and equipment. These seldom present a problem to the larger firms who usually have in-house staff for each of these areas and apply the same thinking and controls to these specialties as they do to all of the areas of the building. But a problem arises when a small firm becomes involved in a project that requires specialized engineering and design. In these cases, get yourself a good consultant to work with you and bring him or her in early so that you form a team from the beginning. You can then present him or her to the client at interviews.

Constructing methods and options are the subject of owner interest, and your firm must support you with background on all the options available as well as the advantages and disadvantages of each in relation to the prospective project. Your firm and staff must be able to demonstrate exactly how they can provide each of the contracting options and how it will affect the professional contract and schedule as well as the construction contract and schedule.

Owners today are concerned with two key factors in construction. First, how can they reduce costs?; second, how can they get their building faster? Hence the birth of phased construction, single and multiple contracts, construction management and turnkey jobs. It is not the purpose of this book to discuss the merits of each, but only to say that you and your staff must accept these variations or come up with your own plans to meet these demands. If you involve your staff in these details in the prospecting period and you are selling the owner a method of contracting new to your firm, your staff must formulate all the details with special care to make it work. Otherwise, you might find yourself with a signed contract that your firm cannot make work.

The owner will always ask, "How much will it cost to prepare for a bond vote or develop the project sufficiently for a firm budget?" The answer once was that the cost is determined by the standard contract for preliminary services. But owners usually felt that the price was too high in view of the risk that the job may eventually exceed the budget or the bond vote be rejected by the voters. The questions for you become: What is the minimum amount of work the firm will have to do to give the owner a design that meets his or her needs? What is a valid budget that will allow the building to be built? Your decision on what to charge the owner will mean the difference between your making a profit or losing a substantial amount of money. I am definitely against giving the owner this type of information without charge. You and your firm must know your costs so that you can give the owner a fair and realistic price for this preliminary work, which, at the minimum, will involve preparing detailed budgets with all items specifically called out and all contingencies noted. Schedules should be detailed step by step, showing the submission and approval process for each stage of the work and the preparation of all documents, including the specific deadlines for each of the owner's activities. Sketches must be included to prove to the owner and state agencies that the building has the required number of square feet and functions as programmed, plus a schematic for engineering services to insure that when the owner says "I approve your budget, schedule and design; now do what you tell me you can do for the price and schedule just quoted," you can deliver what you promised. Your firm must be involved in this preliminary work, which is a vital part of the process of securing a contract.

In all these categories, have your staff research and document the detailed information for your direction. Together, you decide on the action necessary for each prospective job. Your staff can then do the production work to convert the information to usable documentation, freeing you to concentrate on the actual presentation.

Remember your production department includes a highly trained and experienced group of people whose combined knowledge represents a wealth of information in a wide variety of disciplines and building types. Accept them as vital members of the information-gathering team; use them as part of the decision-making process, and your job will be easier and run more smoothly because your decisions will be based on well-documented information.

Department heads are probably the most knowledgeable people in the firm. If you are trying to close a contract with a client who has experienced a great deal of trouble in the area of mechanical and electrical coordination in his previous building projects, you should take your mechanical department head to meet the prospective client. They can discuss the specific problems and your office's ability to avoid such pitfalls. By addressing yourself to a specific problem area and by taking someone to meet the client who can reinforce you in this area, you will demonstrate the capability of your firm to handle this client's work.

A more specific discussion on how each specific department in a firm can help in prospecting follows:

1. SPECIFICATION DEPARTMENT. In a top firm, this is a particularly valuable department. A good specifications writer is much more than a cut and paste person who takes an old specification apart and reworks it into a new one. This person is a materials analyst and a legal and technical writer who knows all the technical details of product manufacture, potential, uses, and installation techniques and can put this information in a coherent, concise, and correct written form. You should sit down with your specifications writer periodically to be briefed on new materials, their advantages and disadvantages as well as the implications of new contracting methods on schedules and costs. This kind of information will greatly improve your credibility.

2. ESTIMATING DEPARTMENT. Your firm's estimator is responsible for keeping records of costs of all your past work as well as the work of other firms in the area. This person can help you keep current on the kind of rule-of-thumb estimates that you must use to answer the owner's questions during the prospecting process. Your estimator must keep you posted on trends and budget projections in relation to escalation. This person can also update prices on your jobs to take account of current price increases.

3. DESIGN DEPARTMENT. Designers often have in their files presentation material prepared in the past for other managers and other projects. It is worth studying this material as well as the projects they are currently working on. Often they are working on a presentation that might relate to the prospective client's project and you could illustrate your presentation with a similar solution that is actually in process. Good presentations always interest a prospective client for they stimulate him or her to think along similar lines about his or her own job. The client may start to create his or her project incorporating your ideas and techniques. This commits the client to you and to using your work as his or her criteria for judging other architects.

4. MODEL SHOP. If your firm uses models in the design process, use them also in searching for new work. Go into the storeroom and see if there are models that relate to the type of project you are currently prospecting. Since much time and money has gone into their preparation, it is good if they can have a new use, and they can usually be cleaned up to look presentable. If you bring prospective clients into your office, they usually spend more time looking at and relating to models than they do drawings and renderings. This is also true if you take models to the interview. Most people cannot visualize three dimensions, and the models help them project their thinking to actual building forms.

5. INTERIOR DESIGN DEPARTMENT. This department is constantly producing color charts, furniture, and material selection boards for present owners to review and approve. You could take some of these presentations to show at the interview. They will attract and hold the audience's attention and help them to remember your firm.

6. PRODUCTION DEPARTMENT. A valuable tool to add to your briefcase for prospecting and interviews are design development drawings as well as working drawings for projects similar to those under discussion. When people see these drawings for the first time, they are impressed and slightly overwhelmed by the amount of detail and volume of material in them. If you leaf through them page by page and tell them a little about each phase of the engineering and architectural processes, you will prove your firm has competence and talent in all disciplines necessary to complete a project. If you add a set of specifications, you

will truly impress your audience. But run through them fairly quickly or you will lose their attention. If possible, leave a set of working drawings with them to review in detail at their leisure.

Behind every successful person is a staff who made him or her look good, but the team concept has been slow to receive full adoption in offices, perhaps because architects and engineers have a strong feeling of individual ownership about their projects.

As you work with your staff and develop them into a support team, constantly be aware of their feelings and try to make them feel important. Never belittle an employee in front of other staff members or behind his or her back. If you have a problem with someone, sit down with that person and talk it out. If you cannot work it out, then you will have to see about replacing him or her with another staff member with whom you can work. If you find yourself in constant conflict with a staff member or in unending discussions about responsibility and communication, you will probably have to replace that person.

You must impress your staff members that their main job is to make your firm look good in the eyes of present and prospective clients.

When you are having material prepared for use at a meeting with an existing owner or prospective client, the number one rule is that you yourself must know in detail the material you are going to present. You have asked your firm to complete the design drawings and the estimates for the projects you are working on. The last time you sat down with them was the week before, when the drawings were 50% complete. You are late getting back from your morning meeting. You rush into the office, grab the tightly wrapped roll of drawings on your desk, and head off to your afternoon meeting. The first time you see the material you are going to present is when you unroll it in front of the owner. This is unfortunate. Sometimes the drawings and material do not even resemble the project you thought was being produced. You must not leave anything to chance. You must know all the details and what is and is not on the final set of drawings you are taking with you. You must also be familiar with the background of the material. You must know the conclusion reached in the report. You must agree with the material and believe in it as the right answer, which means you must constantly review it with your staff during development.

Other Managers Can Help

For some reason, managers in a firm are often made to feel they are in competition with each other, and in firms where several managers are working on different projects, you often have what might be called several firms under the same roof. This is a waste of talent. The other managers can give you a wealth of information on the kinds of projects in which you are inexperienced. If you are trying to close a contract for a building type that is new to you, but which another manager in the firm has had experience, sit down with him or her and draw on that person's knowledge. This manager can give you the working vocabulary that will allow you to talk knowledgeably with your prospective client. He or she can give you programs that were developed for similar projects and can show you where to find recent examples of this kind of work. He or she can help you set up and organize meaningful visits and sometimes attend as the expert from your staff.

Other managers in your firm can also help you with references. In the early stages of prospecting, it is important to establish credibility with the prospective client. Failure to do this will probably mean that your firm is eliminated before the interview. Use the other managers of your office as a sounding board to provide a background in areas in which you lack knowledge. Seek their advice on how to handle the potential client as well as how to handle the project if you are awarded it.

The other managers can also help you with the red tape of obtaining government approvals. If they have worked on several projects with these agencies and this is your first, they can certainly give you some pointers on where to start. Seek their advice in all that you do. Do not copy their actions and techniques, but do build on them to develop a natural expertise of your own.

CHAPTER SIX

GETTING STARTED IN A NEW AREA

Members of a firm often say that since there is no work in their area, they must go to another area, or even another state, to find work. They then go out after a job in a town where they have never been before. If you go after such work without careful thought about where you are going or what you will do if you get the job, you are courting disaster. Before you expand your territory ask yourself how you will serve the clients and whether you can make a profit on distant projects. Ask yourself also, "Have I exhausted every possible prospect in my present area? Have I visited every church, club, hospital, school, and other organization?" If the answers to these questions are no, you have not yet covered your own area. Going out and getting a project farther away from home without proper planning will be a costly, time-consuming operation and may produce a low profit or return on the investment. So before you start out to make the world your territory, stop, think, and plan. Where do you want to go? What kind and size of work are you after and how are you going to treat it profitably for your firm and provide the client with full and satisfactory service?

YOUR TERRITORY

Define Your Geographic Area

Once you have decided to expand your territory, however, and have done the preliminary research, procrastination cannot be tolerated. Until your expansion efforts have borne considerable fruit, you will not have any free time.

First, get a large-scale map of the region in which you plan to expand your practice, and hang it on the wall where you can study it constantly—

as you sit at your desk, talk on the phone and work. You must learn it so well that you can visualize the entire map, identifying all the major towns, rivers, and highways, and especially natural centers that contain airports, interstate highways, and major cities. These centers should be marked with large colored pins. You should then study the map for patterns. Look for the large center that has minor ones within a two-hour drive. This is the population mass you should investigate first. Usually you can fly into the larger center, rent a car, and have a fairly good concentration of population and business potential within a two-hour drive.

How to Service

When you have defined all the centers you wish to consider, you must analyze each and determine which are within an economic traveling distance from your office. Can you drive there, or will you have to fly? What kind of airline service can you use or will you have to rely on general aviation? What is the actual distance in miles? Can you go out for a three-day trip, cover two or three centers, and be home the third evening? How many overnight visits will be required if you obtain work in the area? What are your travel costs in both man-hours and dollars? If you obtain a project and you have to send someone, or go yourself once a week, what will be the cost for the total project? Relate the travel cost to the fee that you will be receiving. Is there enough left over to do the work and still give you a profit? Will the travel take up more than its share of the fee? It is imperative that you consider all these questions.

You must also consider the time element. If you obtain several projects, how many nights will you have to spend away from home to service them? What will be the effect on your other projects? What will be the effect of your being away from your office and your family? These are no small considerations.

But if you review the advantages and disadvantages and still feel that work is out there that you want and can get and that your expenses will be such that you can produce a fair profit, then you must decide how you will service these new centers of business. If they are within a 150- to 200-mile radius of your home office, then it will be a matter of a drive of a few hours before starting work. But if the center is beyond this distance, you must consider how to service the area as a whole. Use of an automobile will be less profitable because it will take too much time, and you must consider air travel. In my experience, medium density areas have proved the most fruitful because there is less competition and because it is easier to establish a presence than in a too dense or too scattered region. Having defined the centers you want to begin with, get a new large-scale map of these areas and develop a full understanding and a complete knowledge of their potential in relation to the type of service you propose to provide.

Potential Clients

In a new area outside your own state, the first thing to do is to visit the state capitol and talk to as many state officials as you can. Do not be put off if they tend to be negative, but listen closely to what they tell you. You will learn much. Then visit various areas of the state and find out for yourself whether any work exists.

Be sure to visit state agencies, including education, state colleges and universities; development, planning, and public health facilities; hospitals and nursing homes; and code enforcement groups. But do not immediately ask for names of potential clients, rather adopt a low-keyed approach. Visit the departments as an interested visitor. Introduce yourself, and bring materials with you to make a short presentation to whomever you can interest in your firm. Tell them about the type of work you have done, stressing any unique projects. Then say that you are considering practicing in this state and wanted to meet the state officials and learn the methods of doing business, codes, approval processes, the general customs and rules or regulations. This will give the official an opportunity to sell himself or herself to you just as you did to him or her. The person may go out of his or her way to give you all the available information. He or she might even give you copies of the latest rules and regulations. You are now on your way.

When visiting owners, you will be familiar with all the agencies, the officials, and the regulations, and this will give you an advantage over the competition. Once an official has given you all the information possible regarding the rules and regulations, you should feel that you can get along together quite well. This is the time to ask him or her about specific owners who might be in need of the type of services you are prepared to provide. The official may give you one or two names of likely prospects. These names must then go to the top of your list of contacts to visit when you start the actual prospecting.

Next, obtain from each state agency a listing of all the facilities within its jurisdiction in the state. For example, the education department will have a listing of all the school districts within the state and the name and address of the administrative persons in charge of the districts. The hospital and nursing home agency will have a similar list, as will the state university system. These are good lists to start with and must be studied for potential future construction needs. Then contact all likely clients to let them know about your firm. This is an obvious practice and all architects will be doing it, but by organizing your list and attacking it systematically, you will be calling on the valid contacts at predetermined intervals, rather than, as many firms do, making a one-shot journey into the new territory, getting busy elsewhere, and never coming back.

The first place to look for the not-so-obvious client contact is in the yellow pages. Get a copy of the telephone directory for each of the new areas in which you intend to practice, and read through each alphabetical listing, asking yourself the following questions. What kind of business is this? Is it one that might be in need of an architect? For example, under "A," we have "Automobile." This is a listing of all the local dealers. In almost every town, a new showroom and maintenance facility is built each year. Pick an agency and then visit with the owner to find out if anyone is building or planning to build in the near future. This same logic should be applied to banks, churches, department stores, golf courses. Visit the largest contractors and get a feeling for who might be in need of professional services. Find out the largest bank in the area and visit the chief loan officer to find out who is planning new construction. Also stop and talk to the larger real estate offices as well as all the manufacturers in the area. Once you have all these lists, obtain a large-scale map of the city and list and locate the potential clients on the map by sections. Then arrange a schedule that will allow you to make the most contacts in a day with the minimum amount of driving. At the same time, you should prepare a written report listing each possible prospect, by type of business, and noting how you found the lead, the names of the person to contact, the size of the operation as well as its potential for expansion or need for architectural or engineering services.

Determine Growth Potential

The next question is which center to start in and devote the most time to. To decide this, you must determine which has the most growth potential in the next five years, for this is where you must concentrate the majority of your action, remembering that construction spurs construction. When one bank builds a new office, all the other banks in town are likely to follow suit. While you are at the state capitol visit the state planning office; get acquainted with the planners and their master plans for transportation, development, land use, and any other programs they might have. This should give you a good picture of population, economic, transportation, and industry trends in the state. But study these plans in the context of past records and present activities. Often the state planners are, as they should be, looking far into the future. But you need work today, so be aware of their future plans but concentrate on what is immediately possible. As you are talking to the planners, search for the kind of information that is not on the maps. Which towns are growing? Which are losing population? If the population is shifting, where is it going? What does all this mean in relation to construction? Is there a boom area for new growth or for renovation and reuse? Are there any key businesses moving into the area? Although these may already have architects or engineers, think about the support services that will be required. If a large industry is moving into town, it will bring people with it who will need houses, stores, and schools.

Make your own assessment of the materials you receive and especially note if present activities in the state are not following the pattern of the master plan. While you are in the state capitol, pay a second visit to the education department to discuss student enrollment patterns. Today student growth is generally related to population shifts, rather than population explosion. In areas where this growth occurs, there will be corresponding growth in services and considerable opportunity for architectural services.

Determine Market Potential

With a background of theory and statistics now at your command, you must now go out and test the market. Spend a day or two in the center that looks the most promising, arriving the night before you plan to get started. Stay in a good hotel in a central location and listen to the conversation around you. After dinner, drive around the center and locate all the key places you intend to visit in the next couple of days and make notes about where things are happening. Who is building new buildings? Who are the contractors? Who are the architects? Who are the engineers? Are they local or from out-of-town? Does one firm have all of the work, or does it seem pretty well distributed among several firms? This observation will give you an idea of the competition. Include the major contractors' offices among your early stops in the process of looking for work. Tell them you are planning to work in the area and want to become familiar with the leading contractors so that when you design and bid your projects, you will have some local knowledge of customs and techniques. During these discussions tell them about your firm and the type of work you are looking for; they may well tell you of prospects, because they will want the opportunity to bid on your work. Be friendly and businesslike, and you will learn a lot.

You should begin your morning calls at the local school district office. If this is a large district, talk to the person in charge of construction who will have a wealth of knowledge about the activities and growth potential of the area. If you talk about your firm and indicate an interest in school work in the area, this person may respond by telling you about pending school construction. Your next stop should be in the local government office. This is strictly a fact-finding visit, and you should not say you are looking for local government work. Do make it clear though that you are trying to find out what is happening in the area and where the future action will be so that you can prospect potential clients.

The intent of this first visit is to get the government representative to relax and talk freely to you about projects of interest. After you get to know each other, the opportunity will generally present itself for your firm to be considered for local governmental contracts. However, if you make this the overt reason for your early visits, you may cause the official to keep to his initial preference for local professionals. While in the local government office, stop to visit the mayor's or manager's office, the planning department, and the city engineer's and building inspector's offices. From all of these visits, you are likely to emerge with several leads, or at the very least with considerable local knowledge, which will help you as you meet other people in the area.

Examine your notes and identify three owners to visit in the afternoon. Remember the basic rule of prospecting. When you make these calls, be sure to ask each owner if he knows of any other owners in the area who might be interested in hiring an architect or engineer. This will probably produce another lead. On these first visits, the basic rule is to ask a lot of questions and listen carefully to the answers. You will learn more than you may realize.

Read the local paper in the evening and be alert to articles that might indicate planning for a new building. Study your morning notes and answer the following questions:

1. Is there a need?

2. What kind of need is it?

3. What sizes are the typical projects?

4. What is the schedule for the upcoming work?

5. Is work here too small, the wrong type, or too large for your firm?

6. If the work is the right type, what are your chances of getting your share of it?

7. Who is the competition?

From this analysis session, you should develop several substantial leads that should be visited on your second day. If you have extra time, also call on the local hospitals, large manufacturing plants, and any other prospects you observe as you drive around.

Analyze Your Results

After you complete your second day's prospecting, you must review your total results and decide whether there really is work in the area for your firm and if you have a chance of getting it. As you rate the areas you visit, it is just as important to consider your potential for success as it is to consider the overall work potential. If you find a good area, know that others are going to find it, too. Sometimes, by letting them compete for a few jobs in one area, you can slip quietly into another and pick up work without competition. When we were trying to expand our practice, we looked seriously at the Boston area and tried to win some contracts there, because there was a lot of construction going on. But the competition was overwhelming; so we looked instead to a rural county in the northernmost region of Maine. There we got more than our share of the work, without competition.

ATTACKING YOUR NEW TERRITORY

Become Involved

The first step toward your goal of work in a new area is to become involved with the place. Your secretary should obtain copies of all the phone books serving the area, and you might consider listing your firm in the yellow pages. Subscribe to the major newspapers, have your secretary read them each day and clip all the items he or she feels would be of interest to you. These would include those about projects in design or construction, pending, or in the committee stage, about companies in the process of buying new property or requesting zoning changes, and items about new board members, the new president of the local bank, and the like. You can quickly scan these and pick out the ones that suggest some kind of action on your part.

You might become active in the business organizations such as Rotary or Kiwanis. If you belong to a chapter in your hometown, you will receive visiting privileges when you are in other towns. This is an ideal way to get to know people who can help you get established in a new community. Another organization you should consider joining is the local business club. Just about every town has one. You can go there for lunch, and it is generally a good place to entertain clients.

When you move into a new area, you should become involved in the local professional chapters as soon as possible. This would include the American Institute of Architects, Contractors Society, National Society of

Professional Engineers and others. It is of vital importance that you help support the organizations. By attending the local chapter meetings in your new area, you will get to know your competition and also what is going on. But do not become over involved, for they will quickly demand too much of your time. Avoid running for office in any of the organizations, at least in the early stages of your involvement, since the time commitment is considerable.

Although trade shows and conventions where manufacturers exhibit their products are often thought of as extended cocktail parties, it is vital that you attend them to find out what is going on and how your projects rate with the competition. Go to these meetings with the purpose of working and finding out firsthand about new products in the field.

Establish a Base of Operations

Your new area or region of work will cause you to be away from your home and office a great deal of the time. In the early stages of prospecting, it will be necessary for you to spend two days or more a week in one of your new centers. You must make a concerted effort for at least six to eight months to see results. It is of the utmost importance that you establish an operating center in each of your new areas of prospecting. This center can be the hotel or motel where you always stay. It is good to establish a routine, always staying at the same hotel on each trip. They get to know you and will generally take messages and hold them for you if they know that you are coming. Also, if you always stay at the same hotel, it is easier for people to find you when they know you are in the area. Another benefit is that during its busy season, the hotel will give you preferential treatment if you are a steady customer. Your operations center can be the conference room at the airport, a local restaurant where they know you and will help you, or perhaps you can use an existing client's office.

Since you will be away from your home and office and must be able to stay in touch and in control, whichever place you choose as your headquarters must have a receptive staff who will let you use their services, will take and deliver all your messages to you, and provide you with a desk or table, chair and access to a telephone.

Making Calls

When you plan your day, group your calls so that you will have as little travel as possible, leaving the maximum amount of time to spend with prospective clients. Another good rule is to group your calls by type. Spend the morning prospecting schools and universities and the afternoon visiting hospitals, other medical facilities, or government and local offices. In this way, it is easier to build your presentation and perfect it as you go along. Establish goals for the day and the trip. How many calls do you want to make? What kind of calls do you want to make? What results do you want the calls to achieve? If you do not have goals, you may find yourself wasting time without accomplishing anything.

Try to schedule four calls per day as a minimum, but six or eight will give you a really full working day. This is not difficult if you are able to schedule your first appointment by 8:00 A.M. or so, and make your long drives during the lunch hour.

On each call, even if it turns out to be a real prospect, ask for additional leads to other owners who might be planning work. Do not push too strongly, but if the person you are talking to seems eager to help further, ask him or her to call ahead to introduce you to his or her lead. Even if this is not possible, a quick question or two before the subject changes may accomplish the same end. Ask about the background of the proposed project and if your contact knows whether an architect or engineer has been hired. The response may be, "I don't know, but let me check and maybe I can save you a trip." This person then picks up the phone, calls the prospect, and finds out for you. If it is a valid prospect, you have an introduction and can be on your way. You must follow up leads as quickly as possible to find out if they are valid. This can be an exciting and unpredictable process, but you cannot afford to ignore leads. When you make a call, do not be afraid to say that it is on Mr. X's recommendation. This is a good introduction and may set the tone for the rest of the meeting. There will be more invalid, than valid leads and as soon as you determine that a prospect is a poor one, return to your regularly scheduled prospecting. But make a point of letting the person who gave you the lead know the results. Even if the results are negative, make the visit a positive and grateful one.

Common Fears and How to Overcome Them

You have done all your research and have made your decision about who you want to visit and which prospects you are after. You drive down the street to the office of your first call. But as you approach the office, you may just drive past the building, finding many reasons for not making that first call of the day in your new territory. This is new work and an unfamiliar area. In other words, you have developed a bad case of jitters, but you must control your actions. The best way to overcome fear is by developing confidence in the value of what you are doing by creating an aggressive, positive action campaign. The longer you put that first call off, the harder it will be to make it. Once you have made your initial contacts in the new area of business development, the entire process becomes familiar and natural, and any misapprehensions will be replaced by excitement and anticipation.

There are five common misapprehensions about initial interviews.

1. OPENING THE DOOR. This, surprisingly is the most difficult thing to do, just like picking up the phone and making a call to a stranger. The only solution is to prepare what you want to say. Assuming a confident bearing, take a deep breath, count to ten, walk up, open the door, and go in. Once you physically touch the doorknob, the rest will come naturally. If your initial greeting is warm, friendly, and sincere, the response will usually be the same.

2. GETTING PAST THE SECRETARY. This is your first hurdle, since one of the roles of a good secretary is to protect his or her employer from unwanted intrusions. On the other hand, most executives want to see all legitimate visitors. The secretary's job is to screen you, form an opinion of you to be given to his or her boss, if asked, and then find time for you in the boss's schedule. When I was hiring people, I would always ask my secretary's opinion of an applicant, and this gave me a good indication of his or her conduct toward other people in the office. Other employers may well value their secretaries' opinions. Treat the secretary as if he or she is the one you have come to see, sell yourself to him or her, and this person will get you in to see his or her boss, even if that person is busy. If the secretary is impressed with you, he

or she will work with you, and if you win his or her confidence, you will likely be given valuable information.

Secretaries to governors, presidents of large corporations, and mayors of large cities take their protective or screening role very seriously. Treat them as if they were the bosses. Make several visits, present them with a picture of what you can do that other people cannot. Once they feel that you may truly have a service that can fill their needs, they will let you through to their employers.

3. WHOM WILL I TALK TO AND WHAT WILL I SAY? This is a problem that we all face. Some people handle it by telling themselves that since they know their subject well, they will manage the conversation as it develops and think on their feet. Although this approach may result in a smooth and interesting meeting, it will not guarantee that you accomplish the purpose for which you came. Always take a piece of paper and make some notes. What are your reasons for the visit? First, to find out if the person or his or her organization will build anything in the near future and, if so, to get an interview with the people who will make the selection of the professionals for the job. With your goals clearly before you, you can ask questions that will determine details about the project and simultaneously present your firm's qualifications. If you are prepared for the meeting, know your goals, and have your presentation prepared and rehearsed, you will perform well and overcome your fears. But do not overrehearse, for appearing natural and sincere is important.

4. HOW DO I HOLD THE INTEREST OF AN IMPORTANT AND BUSY PERSON? This is a comon fear that we all experience as we move out of a close-knit group into the wider business world. We feel secure in our own office and with established clients, but calling on the president of a large corporation, the governor of the state, or the president of a large bank may overpower us by virtue of their public position. A person of this sort is often easier to talk to than his or her subordinate and if you prepare your remarks you should not find doing this difficult.

Respect his or her time and be productive and as brief as possible. Show that your service can help him or her in unique ways, and remember that people on this level are not necessarily very familiar with architects or engineers but are usually fascinated by the professions.

Be respectful, but reveal your pride in your own firm and its work. People at top management levels are usually good at putting others at ease and in quickly discerning their message; so they will usually help you in nervous moments.

5. FEAR OF REJECTION. This is common, but should be put aside, since people usually will indicate clearly, whether you should return for another visit without embarrassing you or making you feel uncomfortable. Of course, there is always an exception to every rule. In the six years that I did prospecting for my firm, I met only one owner who talked to me in very negative terms and viciously attacked architects. The only way to handle this kind of situation is to make a noncommittal response and quickly excuse yourself. But this type of interview is the exception rather than the rule.

Setting Up and Conducting the Initial Meetings

"Should I make a cold call or should I make an appointment?" is a question on which everyone has a different view. Some people say that if you call for an appointment and the client says no, you have closed the door and will never get in again. Another view is that if you call for an appointment and receive an affirmative response, you have given your prospective client the opportunity to prepare a case against you if he or she does not really want to see you. On the other hand, if you arrive without warning, you can surprise the person, catch him or her offguard and get your presentation completed before he or she knows what has happened. But conversely, if you arrive unexpectedly, the person you are trying to see may not be there or may be too busy to see you. It is unfair to drop in unexpectedly on a very busy person and demand a large amount of time.

In the early prospecting process, if you do not know who you want to see or where they are, then a cold call is the only way. But you must have a reason for the call. You must make your visit appear natural and not embarrass the prospective client by telling him or her that you went several hundred miles out of your way to see him or her. Tell the person that you were in the area on other business, had a few minutes, and dropped in to get acquainted. This will put the person at ease and indicate that you are doing business in the area. He or she will understand that you are simply introducing yourself. This can almost be done at the front door. It is not necessary to get into the client's office on this first visit. However, if you have your opening statement planned, it may stimulate the person's interest sufficiently for him or her to invite you in or to return another time. You must create interest so that the person will want to make time for you now or later. This is when all the time and money spent on researching the project and the person before your visit and on preparing and rehearsing your opening is put to the test.

Once you are accepted, always be expected. This is the basic rule of making calls. Your time is as valuable as the person's you are visiting. Let him or her know when you are coming to town so your visit can be productive and to the point. The only exception to this rule is when you are simply keeping track of a prospect and may call without a special appointment to find out what is happening and renew your expression of interest. If you try to force or constantly inconvenience someone, he or she will no longer welcome you.

Appropriate here is discussion of telephone sales surveys, sometimes conducted to find out what is happening in an area. I was in an owner's office once when the phone rang. The call appeared to be from an architect looking for work. After the call was completed, the owner commented that the architect in question must really be in trouble because he had his secretary call. It sounded as if her boss had given her a list of questions and she was calling all of the owners in the state to find out where there might be some work. This is not the way to impress an owner. The telephone is a vital tool that can save you much time if used properly. Consider a typical state that has thirty or forty school districts, each of which is a potential project. The question is where to start. It could take a great deal of time to visit all the districts. A quick telephone survey will help you uncover the district that has the greatest potential. Using the phone is like opening the door for the first time. The planning is the same and must be just as carefully carried out. You must know what you want to say and you must have a goal to work toward on the phone. The typical goal is to find out if this district is planning any construction, and if it is, to get yourself considered as a possible candidate. You also want to know if there are other districts or owners known to the person you are calling who might also be planning construction. The difference here is that you are not visible to the person to whom you are talking, so you must develop the technique of conveying excitement in your voice. The person you are talking to must sense that here is a person or firm that is sufficiently interesting to be worth talking with. Learn to talk with your whole voice, not just with words. People relate to how you say something, as much as to what you say. This is the key. Once the person on the other end of the telephone has started talking, listen to what is being said and the tone of voice of the speaker. If the situation looks promising, make an appointment, but be sure you keep it. It is a great mistake to call up twenty districts and promise each one you will stop by, when you will not be able to do so in a reasonable time period.

By the time you have completed the extensive survey, you will have a wealth of information, which must be recorded or you will quickly forget which were the most promising leads. Map and grade the districts in order of potential and record the contact in each case. Then use this information in your prospecting. It is not worth the time and money devoted to these calls if you do not use the information. You must avoid the reputation of a conductor of surveys and gatherer of information and become known as the person of action and the firm that gets work.

CHAPTER SEVEN

THE INTERVIEW PREPARATION

The most important thing to remember is if a job is worth going after, it is worth your time to plan the interview. People often spend a great deal of time and energy trying to get to the interview stage, only to ruin it by not preparing for the presentation. They often make the excuse that since they know the situation and subject so well, they prefer, as they say, to "play the interview by ear." They say they don't want their presentation to sound canned.

Failure to prepare fully for the interview is a quick way to disaster. The interviewing process, as we know it today, is a complex psychological contest. Interviews are not won just on what you say, but on how you say it and how you react to the prospective client and his specific situation. Therefore, it is not only wise, but necessary for you to plan and prepare in depth for the interview.

STEPS IN PLANNING THE INTERVIEW

Define the Projects

Defining the project is the first step in the process of planning. I am sure all of you have been in a situation similar to one I experienced when we were interviewing for a school project in northern Vermont. We were well into our standard slide presentation on school design and construction. In the middle of the slide show, the chairman of the board raised his hand and said, "I have a question. Do you know what we need? All we want to build is a new gym and a new library. You are talking about a whole new school. It would be nice, but we can't afford that." I knew right then, the interview was over for us.

To repeat, be sure you define the project. Is it a completely new school, church, or hospital? Or is it an addition to an existing structure? What size project is it and how much money is budgeted? Plan your interview to show the prospective clients that you can give them exactly what they want at a price they can afford.

After you have defined the project, address yourself to defining the type of service needed. This may vary from a preliminary feasibility study to a full-service architectural and engineering contract. In today's market, more owners are asking for total services that include construction management as part of the architectural and engineering services.

If your office cannot provide the comprehensive services requested, you will have to bring in consultants. If this is the case, select them early so that they can help you in the preparation of your presentation.

After you have defined the project and the services, take a look at the owner's time schedule. Are there any pitfalls in the schedule? Owners often procrastinate. When they do finally make a decision, they are apt to want the building completed and ready to move into very fast. What can this do to you? Suppose you are in the middle of your presentation and are interrupted with this question. "Can you make it possible to have a bond vote in sixty days and guarantee occupancy by next fall?" The wrong answer at this point can cost you the job. If you laugh and say that does not sound possible, you are eliminated. Someone will say yes and will get the job. They might not perform and they might not get the next job, but they will get this one.

How can you avoid this situation? Know, in advance and in detail, what schedule the owner hopes to keep. Work, before the interview, with your staff to find a method to meet the deadlines. List all the steps you must take and deadlines for each. Then list all the steps and deadlines that the owner must meet. Now you have the answer. The owner will be impressed with your thoroughness in planning and identifying the details.

You have now done two things: you have personalized the presentation to his specific project and you have pointed out the problems that face both sides in meeting any schedule. These are points in your favor in the selection process.

The next two things you must emphasize in your presentation are surveys and borings and the money available for this work. Failure to do this can be disastrous.

Owners do not always understand what is meant by survey and borings, nor the time and cost involved and who bears the cost. Since you cannot start work without completing the survey and borings, you must investigate the details and be able to explain them to the owner. Be careful not to overemphasize them and scare the owner, but be sure you emphasize your respective responsibilities in this important, but sometimes neglected, phase of the work.

Before the interview, determine how much money the owner has available for preliminary, as well as for full architectural and engineering services. When is this money going to become available? In today's tight money market, when owners are finding it difficult to finance their projects, we often see a reduction in the money available for preliminary services.

Some owners will even insist that you perform the services necessary for a bond vote without fee. Then if the vote is successful, you become the architect or engineer and will be paid. This brings up a question of ethics, but this is not the important point here, since, in my view, working without fees can never be profitable in the long run. As a businessman who is always interested in a profit, you will find that working without being paid is a sure way to disaster. It would be wiser to spend the same amount of time and money taking care of your present clients and prospecting future work than on working for someone on speculation. When you give free services, the owner does not appreciate you or what you are doing. There is no loyalty from him or her and little incentive to move ahead with the project; consequently, the project is likely to drag and cause your overhead to go up. If the owner does not have anything invested, he is probably not going to work very hard to move the project along or to get funding approved. In short, you have nothing to gain and everything to lose.

Sometime you will be asked to work without fees. You must understand the many implications and prepare a positive approach. The best solution is to show the owner how he can stay within his or her limited budget and still get the necessary services completed for funding approval. You must anticipate this situation and prepare for it.

When an owner says, "Will you do the preliminary work without charge, prior to funding," avoid saying yes or no, since in either case the situation will work against you. Either you will not be hired or you will lose your fee, if the funding is turned down and the project stops. Be prepared. Have a positive option that the owner can accept that will not hurt you financially. These options are limited only by your imagination, but could include reduced services to fit the budget, deferred payments or other options.

Define the Client

Defining the client is the second step in the process of planning. Just as defining the project gave you insight into what the owner wants and how you will propose to provide it, defining the client will give you insight into who he or she is, how this person thinks and what his or her personal needs are. This will help you prepare your presentation and your answers in the way he or she can best understand and relate to.

Who is the client? Is his or her project public, private, or semi-public? Who is the full-time leader for this project? What is his or her relationship to or with the owner? What are their qualifications? What is his or her personality type? What does he or she think of you and your firm? Remember that, while a full-time project leader may not have a vote, this person may have a strong power of veto.

During the prospecting process, you usually make all your visits to the full-time leader. You gear your presentation to what he or she wants because this is the easiest person to find and also because he or she knows what the project entails and speaks your language. Without his or her endorsement, you will not get an interview with the board, the members of which make their decision based largely on this person's input.

With this background, you should now think about the people who will vote. They may be the board of directors or a building committee made up of board members and empowered by the board to interview and select a firm. Be cautious of appointed building committees since they do not always include board members. Sometimes they are not empowered to act. When this happens, you can expect a lot of wasted effort. You will think you are all set and have the interview won only to find this group does not have the authority to make a decision, and you will have to sell yourself all over again to the board. There is no real answer to this except bear with it and keep going. In your preliminary talks with the committee, find out who the leader is. Bring to his or her attention that the study committee does not have official status and discuss what this means in terms of time and total schedule.

As you research the board makeup, determine each member's attitude toward construction. Are they in favor of this project, or are they being forced into it? What is their thinking about construction, modern, traditional, conservative? What are their ages? If they are old, as board members often are, they may have difficulty in hearing. If so, they will appreciate your talking distinctly and using large-scale audiovisual material because they may also have impaired vision. The key presentation should be by a senior member of your firm who has a mature presence and bearing and is clearly a leader in his or her own community.

While you are analyzing the board and deciding how to approach them, determine who the leaders are and who will be the decision makers. This will give you additional background in organizing your presentation to win these key people to your side.

Once you have decided how to organize the presentation and who will make it, there is one last thing to do. Obtain the names and backgrounds of all board members. This background should include where they live and work, their positions on the board and in the community, their hobbies and interests. This information should be typed and taken with you to the interview and reviewed before the meeting. When the opportunity arises to chat with any of the members, you can carry the conversation into his or her area of interest. This will create a favorable impression.

How Knowledgeable Is the Owner?

How much background in construction do the owners have? By owners, I am referring to the board members or others who will be responsible for interviewing you and making the final selection of the professional firm. Ask yourself some of the following questions: Do they know what they want to build? This is sometimes different from the project. How knowledgeable are they about the construction process? Do they understand construction jargon? Are they up-to-date on new trends in construction as well as the new uses of buildings of their type? Do they know what an architect or engineer does? Answers to these questions will show you the owners' level of understanding and allow you to build your interview on this level—not above or below it.

During the interview, owners might ask a number of previously prepared questions. If you have not researched their knowledge of construction, you might be misled. The questions often indicate a deeper understanding than they actually have. Here is a list of typical questions owners usually ask or think they should ask. They do not necessarily understand their own questions or the answers.

1. What size firm do you have?
2. What experience do you have in this type of project?
3. Do your roofs leak?
4. Who will be in charge of our project?
5. What is this person's experience?
6. How long has this person been with you?
7. What happens if he or she leaves?
8. What consultants do you use?
9. How do you control your consultants?
10. How much will our project cost?
11. When can we start construction?
12. When can we occupy the building?
13. What kind of construction do you use?
14. What mechanical systems do you recommend?
15. How do you handle energy conservation?

These questions could indicate a high level of understanding of the construction process, or they could be a list prepared by someone else.

Your answers must be intelligent, correct, and correspond to the owners' level of familiarity with the construction process.

Your research into the board members' background and knowledge will help you uncover the oil salesperson, the roofer or contractor who is sitting quietly waiting to sink the first architect or engineer who incorrectly answers any question in his or her area of expertise. Such people strongly influence board members. A single wrong answer to an oil supplier or other construction-oriented member might be the basis for your elimination.

Defining the client is the most important item involved in prospecting and winning the contract. You must know who you are talking to, the person's background and how knowledgeable he or she is in construction. If you guess wrong, you will either be talking down to them or talking beneath them. In either case, they will probably dismiss you.

Has this owner had a previous building project? What was it? When was it? Who was the architect? Was the project a success? It also will help to find out if any of the present board was involved. This is the kind of information that will help you determine the extent of the board members' knowledge and understanding of construction. If they had problems, find out what they were and how they were solved. If they had a problem on their last project, you can be certain that they will ask you how you would have solved it. Be sure you give the right answer. This type of research will give you a feeling for their relationship with their past architects or engineers, and their attitude toward such professionals in general.

In looking at the board, be realistic and do not cut owners short in conversation. Owners often do know more about construction than we architects and engineers give them credit for.

Acknowledge owners as "members of the team" and accept the fact that they do know something about construction. Owners will have to operate the buildings you designed long after you are gone. You may be able to walk away from your mistakes, but they must live with them. Communicate with them on this level at the interview and throughout the entire project and you should never run short of clients.

Define the Interview

Defining the interview process is the last step in the planning research prior to the interview. If you have done your research and obtained answers to the questions discussed, you will already have a good deal of information. This is the time to go back and review your prospect file and refresh your memory of the prospect.

How was this prospect found? Was it a cold call, a referral, or did they call you? In short, how did they hear about your firm? How much do they know about you? Have they seen any of your buildings? Do they know any of your clients?

Next, look at the actual setting up of the interview. What kind of interview will it be? Is it a screening interview to reduce the list of candidates to a reasonable number? If this is the case, candidates are screened and reduced to a workable number of from three to twelve.

You might be asked to a follow-up or full service interview when the list of candidates has been reduced to two. This interview would be a specific, narrow-scope interview to select the winner. Each of these three types of interviews requires a different approach. The approach you pick can only be tailored to the specific conditions; it cannot be standardized.

How much time have they allowed for the interview? Some owners will try to schedule an architect for a half-hour presentation with four to six firms scheduled for one sitting of the board. This is disastrous. You are apt to talk too fast to cover the material you think important and end up in a race against time. With five or six other firms doing the same thing, it turns into a game of musical chairs with the last firm to make the presentation the likely winner.

If you are caught in this situation, arrange to be interviewed last. In any event, you must resist the urge to tell them everything you know in the half-hour time limit. Your research will be vital at this point. You must find out what they want covered and pick a topic that is simple and understandable. This should also be one that will interest them and make them want to call you back for a follow-up interview.

A good length of time for an interview is forty-five minutes to one hour. This will give you time to make your presentation and allow some time for questions. If more than an hour is scheduled, you will probably bore the audience. The only time you should stay beyond an hour is if they are asking questions and want you to stay. If this happens you are on the way to winning the contract.

Find out which other firms are being interviewed. This will let you know who your competition is and what you need to do to come out on top.

The interviewing process may go on for weeks, and the big question always asked is, "Are we being interviewed last?" Although the odds on being hired are generally better if you are last to be interviewed—and first is better than being in the middle—do not give up just because you are not in either of these favored positions. Concentrate your efforts on doing a thorough job of prospecting and getting to know the people on the board. Then do your research and give them a presentation they will remember.

Develop Your Outline

When your background research is completed, you know what the project is, who the client is, his or her background, and what the interview is all about. It is now time to start developing your outline for the presentation.

You must decide the main points of your presentation, which should be based on the results of your research in terms of the owners' wants and needs, such as performance on a tight schedule, low cost, new ideas, experience in a building type or full in-house service. Pick the ones you feel are most relevant and make notes on how your firm can fulfill these needs. List your strong points for each need, from the most basic to the most important.

As you develop the outline, put your material together to build and hold the interest of the audience. Start with the basic material, and build up to the most exciting so that your high points of excitement will come at the end of the interview. This provides a stimulating climax, and when the audience becomes interested and excited, they will begin to ask questions. This is a sign that you are getting through to them.

At one school interview in northern Maine, we made our presentation according to this method. We forced an ending when the audience had reached a high point of enthusiasm and asked if anyone could provide transportation to a motel, since it was too snowy to attempt to fly home. The superintendent asked us to wait in his office while the board conducted some final business and then he would take us to a motel. In a few minutes he asked us to come into the board room again to answer some extra questions. When we arrived in the room, they had a signed contract on the table and asked us to read it and if we agreed with it, we should sign it, for they were anxious to get their project started. They also asked that we stay over the next day to get things moving. This contract included a new 500-pupil middle school as well as a survey of their other buildings, which led to additions and alterations to most of these buildings. Immediate closings do sometimes happen.

As you develop your outline, be sure to include the following critical items. Weigh them and place them in order of importance.

1. BACKGROUND OF YOUR FIRM. This should include who you are and what your general accomplishments have been. Include the names of your key people and the names and positions of the people who will handle the project. This is also the place to include references to your awards and other noteworthy accomplishments. Keep it interesting and avoid lengthy details and do not be self-congratulatory. Give factual information, supplying details only if asked.

2. TYPES OF SERVICE. Under this heading you should list all the items of work necessary to design and construct their building. This gives you the opportunity to describe the process and refer to the parts of work you will handle in-house. Also list the work for which you do not have in-house capability. This gives you an excellent opportunity to introduce your consultants and describe their background in relation to this project.

3. HOW YOU WILL HANDLE THIS PROJECT. Here you can list the step-by-step process for their project from start to conclusion. Remember, this is only a key-word outline. There should be as much blank paper as there is writing. This will give the prospective client a place to write notes as you use the outline in your presentation. This is also the time to cover the qualifications of the person who is going to be directly responsible for their project.

4. YOUR PLANS FOR COMMUNICATION. Lines of communication to keep the project moving and to keep all parties fully informed should be mentioned at this point. Good continuing communication is the key to success and a happy owner. Never forget that a happy owner means more work and good references.

5. SPECIAL SERVICES TO MEET THEIR NEEDS. This is where you will unveil your surprise. Include here your ideas for this particular project, which you have developed as a result of your research. List special talents to demonstrate your unique qualifications. Create excitement. Remember, this project is a major undertaking for the owner and very likely a once-in-a-lifetime experience. Make it sound exciting and get him or her excited about building. Constructing a building can and should be fun for everyone involved. If it is not, you are doing something wrong, and you will fail to attract repeat clients.

This approach of presenting exciting new ideas and options at the end, will help you build to a high climax. New ideas and options will start them thinking and asking questions. The more questions you can get them to ask, the better your position, especially if these questions relate directly to solving a specific problem within their project. Schedule the last phase of your presentation to elicit this type of response and you will be another step closer to winning the contract.

Assemble Your Material

When you have assembled your material and the items you plan to use, review them and assign priorities to them. If you have one hour for your presentation, assign the time so that you devote the most time to the most important subjects. Typing drafts of your outline may discourage your secretary, because you may have to redo it many times until you get it the way you want it. You should make your secretary a part of the composing team. He or she will then be more enthusiastic about typing the outline repeatedly.

The finished outline must be sharp and simple in appearance. This is the first thing you will give to the people at the meeting, who will be referring to it throughout the presentation. If you use the key word and reference method, the outline will be easy to follow and understand.

Check your outline and the materials you have assembled. Be sure that everything you want to cover is included. As you prepare the material to be used at the interview, make it as personal as possible. Include the name of the owner and reference to his or her project on all handouts. It is vital that all material be carefully proofread for proper grammar and spelling. Misspelling the name of the owner or his or her town or using improper English can irritate him or her and get you off to a bad start. You need everything to be in your favor.

There are many kinds of handouts you can bring to an interview, but the ones most useful are listed here:

1. REFERENCES. These prove that you do what you say. As you decide which references to use, choose those in the same geographical area and of a similar building type. Check your references to be sure that if approached they will give you good recommendations. Avoid listing government officials as references. Once you have decided whom to use, prepare a single page in the same format as your outline, and list all the references complete with names, addresses, position, type of project, and phone number.

2. VISITS. Prepare information on the projects you would like your clients to visit. Be certain that these projects have some application to their work and are in a condition to be shown at an advantage. Prepare a visit handout in a corresponding format. List the name and description of each project as well as the person they should contact at the job. Include a building floor plan, statistics on building type and size, cost as built, updated cost to build today, and any other basic information you think is vital.

3. CONTRACTS. Owners will ask you what kind of contract you suggest for their project. Architects typically answer, "We use the standard AIA contract." But the question is a loaded one and will cause many problems in the future.

   There is a standard AIA form that you can use, but each architect and engineer must modify this form to suit his or her own method of operation. Thus owners may think they are getting one thing, until they get the bill for services, or the architect tells the owners that certain services are not included in the scope of service and will carry an additional cost. This causes hard feelings, which may quickly cause the relationship to deteriorate.

   The best way to handle this is to prepare a contract, based on the services that you propose to provide and including all of the variation from the standard. The contract must reflect the services you are selling at the interview and should be signed by the partner to signify that you are willing to do what you have told them in the interview you will do. But be careful how you present such a contract or some board members may interpret it as coercion, to try to get them to sign. This is not the intent. The signed, sample contract is additional documentation to show the owner that you stand behind and can carry out what you are proposing at the interview. An option you should consider for smaller projects—if, for example, the owner only wants you to do a site selection study—is to prepare a one- or two-page typed proposal listing the scope of the work, the time limit, the proposed fee, and when it is to be paid, as well as the material and services to be furnished by the owner. This information can almost always be put on one page. This short-form proposal must be prepared ahead of time and presented for discussion at the interview. Whatever you do, do not leave this question regarding contracts unanswered.

4. FEE CURVES. Inevitably fee curves will be the subject of discussion and can create considerable confusion if not handled properly. Most owners at the time of interviews do not know the exact value of their projects. Moreover, there may be confusion over the definition of new work and alteration of the existing structure. Your office should have a standard curve for each building type. Each curve would have two lines on it, one for new work and one for alterations. Your firm's definition of what constitutes new work and what are alterations should be written on these curves. This should be simple to read and understand and should reflect an acceptable profit margin for your firm. Attach a cross-reference to these curves to your contracts. If as usually happens, the scope of the project changes, you are protected and the owner has been advised in advance. He or she is not surprised and everyone is happy.

5. THE OFFICE ROSTER. This is of value if you are selling the size and quality of your staff. Using your standard format, you can prepare a listing giving the qualifications and experience of partners and key staff in each department of your firm.

6. SPECIAL INSERTS. These will cover all the other items you might want to include for an interview, such as reprints of published articles, a listing of similar projects completed by the firm, all recent awards the firm has won, and special case studies of noteworthy projects or construction techniques. These should be added to standard handouts, such as your company brochure, and presented as a coordinated package in the order in which you will refer to the items. The cover should include the name of your firm, the name of the prospective client, and the date. This will convince the board that you cared enough to prepare your documentation specifically for them. Avoid overloading the audience with an unmanageable volume of material. Do not assume that all the items listed should be used at each interview. Evaluate each, on the basis of your research, and then decide which will do you the most good.

Travel Plans

Planning your travel is vital, for all members of your team must arrive at the interview with time to spare and with all of their materials. Find out exactly where the interview will take place. "At the school" is not a sufficient description. You must know the room number and the exact street address.

Whom Do You Bring?

This is a difficult question to answer and demands careful thought. If you bring too many people, the audience will become confused. If you bring someone who does not usually meet the public, this person may do you more harm than good. The following points should be remembered when assembling your supporting team.

1. KEEP THE NUMBER OF PEOPLE AS SMALL AS POSSIBLE. If you have a one-hour interview and you bring six people, the names and faces will change each ten minutes. This confuses the audience. In a one-hour interview, keep the number of speakers to three or less and never more than four.

2. IF YOU INVITE SOMEONE TO COME, GIVE THAT PERSON A REAL PART. When you bring someone new from your office to an interview, this person may easily be overpowered by the whole process. Remember this is a new experience for him or her and, although as the leader you might expect to dominate the situation, do not neglect to use your other members. If after an interview, someone says to you, "Who is that other person with you?" or "We wish we had more time to listen to the rest of your team," take note. You have been guilty of running a one-man show. Brief your team members on what is expected of them from the minute they enter the room to the time they leave. Give them a real part in the presentation and help them rehearse it. You are the captain of the team, but you cannot win the game by yourself. Proper selection and training of your teammates will do a great deal to make yours the winning team.

3. BRING IN THE EXPERTS. During your research and development of background information on the project, you will uncover several things. You will establish any unusual requirements or areas in which you are not fully familiar. Accept your limitations and bring in your experts to cover those areas you feel will be brought up during the interview. It is not a disgrace to admit you do not know the answer but it will be a disaster if you try to answer something about which you know nothing.

4. YOU MUST BRING ONE OF THE FULL PARTNERS OF THE FIRM. The owners want to know the person responsible for signing the contract if they hire you. They also feel that if the full partners are not interested in attending the interview, they will not be very interested in being selected. Professional firms have a high rate of turnover, and project managers who start the job, sometimes leave the firm before the project is completed. Therefore, the person who will ultimately be responsible for the project and in binding the firm to perform, must be present to meet the owner.

5. HAVE THE PROJECT MANAGER WHO WILL BE RESPONSIBLE FOR THIS PROJECT PRESENT. Make the assumption that the top three firms being interviewed have equal technical skills. The success of any project rests on the personality and capability of the project manager who will direct the work and keep it moving. For this reason, the owners will

want to meet the manager and determine how well they will work together. It is important that project managers be people who like and get along well with others.

THE REHEARSAL

Dividing Up the Parts

A smooth interview is one that has been well planned and well rehearsed to reach a level of perfection that will allow you to appear polished and knowledgeable. The rehearsal will help you to cover the maximum amount of material efficiently in the time allowed. It will also give you practice in answering unexpected questions and in becoming proficient in responding to questions under pressure.

Divide your presentation into parts. Appoint one leader, and give all the other people who will be attending the interview, equal time. As you divide the presentation, keep in mind the transition from one speaker to the next. Changing speakers too often will confuse your clients. If you have a person scheduled to speak twice, can his two presentations be consolidated? The leader, of course, speaks at several different times. He or she provides the needed continuity. All others should have one chance to do their entire presentation and then take an active part in the question and answer segment at the end.

As you assign parts, determine the extent and type of audiovisual material to be used. If one audiovisual item can replace a speech, use it. People remember pictures better than words. Determine the amount of time each person will require to make his or her presentation and relate that to the total available. Use only the number of slides needed to tell your story well without repetition. Too many may put your audience to sleep.

In a forty-five minute presentation, try to allow at least ten minutes at the end for questions and answers. Be prepared to fill this void if your material does not initiate questions. But your material must generate questions. Rework it until you are sure that there will be questions asked.

As mentioned earlier, review the importance of each item in your outline as it relates to the needs of the audience. Assign time accordingly. Give the items of most importance the most amount of time. The speaker covering the most important segment of the presentation should speak last. He or she must have the ability to generate enthusiasm. He or she must be able to bring the audience to the high point at the end of the presentation. If not, you will be forgotten as soon as you leave the room. Everyone else should build his or her material to complement the last speaker and to lead into and support the next speaker's presentation.

Key-Word Notes

Key-word notes should be developed by each speaker for his or her presentation. These can be placed on the outline in the blank space left for notes. If you prefer a more organized method, they may be kept on a separate piece of paper. In no case should the paper be a full 8½ × 11. This size paper with notes, scares the audience. It indicates you are going to read the material and that you have a large amount of written material to cover. Written material puts audiences to sleep. Fold the 8½ × 11 paper into thirds. This is adequate for key-word notes.

Your key-word notes should list items and statements in the order in which you plan to use them. During the presentation you can refer to the notes to make sure all items have been covered. At the end of your presentation, it is an acceptable technique to say, "Before closing, let me check my notes to be certain I have covered all the material." This provides a good summing-up device, which will also give the audience a chance to think before you sit down. It is a workable, built-in pause, allowing people time to think about their questions.

Miscellaneous key-word notes might consist of questions you would like the audience to raise, terminology related to this type of project. Make a note on your outline of the number of minutes you have been allowed for your section. As you start, note the exact time and add the time of the presentation to it. At the top of the page, write in large numbers, the exact time your presentation must end. By placing your watch on the table next to your outline, you can keep track of your time and end on schedule. Insist that all speakers do the same thing and you will cover all the material in the time allotted and still have time for questions and answers at the end.

Signals

Signals between the person talking and the person on your team who is listening will alert the speaker to respond quickly to any audience reaction. How many times have you been to a meeting where someone gets up to talk who will not sit down? They become so involved in what they are saying, that contact is lost with the audience; thus communication is impaired. When you make a presentation, assign one person to listen and observe. Agree on signals to use when you seem to be losing the audience for one reason or another. Listed are some of the signals we found useful.

1. SPEAK LOUDER. This is required when you observe the audience straining to hear the speaker. Many owners and board members over fifty years of age have hearing problems. Do not be afraid to signal your teammate to talk more loudly. A simple signal is the hand cupped behind the ear. Your audience will go to sleep if they cannot hear you.

2. MORE DETAILS. These should be called for if you see that your teammate is discussing a point to which the audience is relating positively. An effective signal to encourage him to go on is a smile and nod. If the audience sees this, they will accept it as an expression of your approval and enthusiasm.

3. FEWER DETAILS. Give such when your teammate is talking a subject to death. This may happen when he or she gets too involved in an answer and wants to tell them everything he or she knows. If you see the audience getting restless, give the sign to move on to another subject. A good one here is to slightly scratch your head and frown.

4. CUT IT SHORT. Do this when the speaker is running over the time limit or does not really have the audience's attention. To give this signal, just run your hand across your throat. Use it when the speaker is seriously cutting into the next speaker's time or when you feel the audience is not interested in the subject. This is no reflection on the speaker, but rather a flaw in your planning research prior to the interview. Move quickly to the next subject. Recapture your audience.

5. SIT DOWN. The signal for this should never have to be given if you and your teammates have really rehearsed your interview. But sometimes a teammate thinks his or her remarks are impressing the listeners, but from where you sit you can see the audience's reaction which tells you the reverse is the case. Get the person off the floor at once. Usually a frown and the "cut it short" signal is sufficient. If not, and you are in a position to move around, walk behind the audience and give him or her the signal again. If he or she misses this time, involve yourself in the presentation. Casually take it over, then move on to the next subject. After the presentation, review the problems with your team, so that it never happens again.

Audiovisual Support

Good audiovisual material is a vital support element for your presentation. The old saying, a picture is worth a thousand words, is especially true in an interview. If you focus your audiovisual material on the specific owner and include detailed references to his or her project, your pictures will convey a great many thoughts in a few minutes.

If the owner has given you a short time period in which to make your presentation, audiovisual material can help you include all the necessary material and keep you in the owner's mind long after the interview is over.

There are some basic rules to follow in preparing your audiovisual material regardless of the kind of presentation you use. Listed here are some suggested key rules in relation to architectural interviews:

1. EACH AUDIOVISUAL ITEM SHOULD BE SPECIFICALLY ADAPTED TO THIS INTERVIEW. Include the name of the owner and reference to his or her project.

2. KEEP THE AUDIOVISUAL PRESENTATION SHORT. We tend to show three slides when one good one would be sufficient. We also tend to show three sketches or other illustrations when one would do. It is easier to hold a person's interest if you change the media often rather than concentrating too much on one type.

3. AVOID THE USE OF SMALL DETAILS AND FINE PRINT IN YOUR PRESENTATION. Use large bold strokes and strong colors. Floor plans are very difficult to read from across the room, especially if they are on slides. Use your imagination and make your presentations readable. Remember once again your typical audience is over fifty and their eyesight may not be perfect. They must be able to read and understand your charts or illustrations from across the room.

4. LARGE GROUP AUDIOVISUAL PRESENTATIONS ARE PARTICULARLY VALUABLE IN COMMUNICATING GENERALIZED MATERIAL AND INTENT. If you want to go deeply into specific and exact details, use handouts each person can hold and read at his or her own pace and on which the person can make notes as you go through your presentation.

Slides show buildings in glowing color; they are produced by professional photographers under ideal conditions that are seldom obtainable. These are the most common tools used in interviews. When an invitation to attend an interview is received, it characteristically causes some professionals to go to files, pull out a large selection of building types and assemble them in a slide tray.

Put yourself in the position of the owners who are listening to the third presentation of the evening. The lights go out and the slides go on. After three sets of slides showing similar building types, combined with speeches on the virtues of each building, in a hot room, at a late hour, the inducement to go to sleep is very strong. Once an audience goes to sleep, you will never alert them again.

How do you overcome this? Look carefully at the slides you plan to use. Reduce the number to twenty. As you look at the slides you have selected, determine exactly what it is you are trying to show. If you have two or three slides that tell the same story, pick the one that shows the best idea. Assemble the slides in an order that tells a story. Too often the typical commentary goes as follows: "This is one of our hospitals. This is the main entrance. This is the patient's room. This is the operating room. This is the administration area. This is another hospital." Instead, try grouping all your hospital exteriors together and saying, "Here are some examples of the exterior designs of our hospitals." Do the same for other parts of the building.

Choose a slide to indicate the beginning, either a title slide or another opening idea, and one to bring the presentation to a clear conclusion. It should be something original and eye-catching that the audience will remember after you are gone.

Once you have your standard slides assembled in an order that tells a story with a clear beginning and end, run through your slides again. Do they tell the story you want to tell? Do they look the same as the slides of others? What can you do to make yours outstanding? Can you add some graphic slides to illustrate some particular points and focus the presentation on the specific audience?

At a meeting in New Hampshire of the New England Hospital Engineers, several architects showed examples of recently-completed hospitals. After the second presentation, all of the hospitals began to look alike. But then, one firm announced they were going to show a case study of one of their hospitals. This consisted of a series of drawings showing the evolution of the space and the decisions necessary to create it. Their graphics led into the slides of the completed building. Three years later, I remember this architect, his name and the building. Why? Because he was different. Tell your story, but tell it in such a way that it is unique.

The remaining audiovisual material is divided into four types: preplanned broad scope, preplanned narrow scope, exhibits of similar projects, and action aids.

Preplanned broad-scope items would be illustration boards or other large presentations giving broad conceptual statements. An example might be a large colorful chart showing the escalation of building costs over the past five years and projecting this into the next five years.

A preplanned broad-scope presentation can consist of anything you wish to discuss in conceptual terms and should be readable and understandable by a layman.

Preplanned narrow-scope exhibits could be detailed worksheets for computing the cost of the project or developing its schedule. These would be presented to each member of the audience on $8\frac{1}{2} \times 11$ sheets and would contain some standard information with spaces for the members of the audience to fill in with specific information on their subjects. Excellent results come from this type of material. It allows audience participation in the development of specific facts in relation to their projects.

Mounted exhibits of similar projects are examples of material often overworked. The professionally made photographs and the large color displays are excellent for national exhibits and other mass viewing, but in the interview situation, they become overpowering and appear too expensive and elaborate. They are also bulky to move in and out of rooms. People have a difficult time visualizing a drawing, but if you put a three-dimensional model in front of them, they will often discuss it animatedly for hours. A take-apart model is even more popular, since people like to move the parts around. The use of models to express design ideas is effective. They can be simple cardboard study models or sophisticated finished models of specific projects. But always ask yourself the question: "Does this presentation material apply to this particular interview?" If so, exploit its full potential, and do not introduce it just for quick reference.

Action aids are vital to your success as a public speaker. Action aids are flip charts, chalkboards, or any other surface upon which you can write or draw to illustrate or emphasize particular points in your discussion as they occur. Ask any teacher how to keep the attention of a class, and they will always tell you, "Don't lecture." Then they will tell you to use a blackboard to write or draw what you are discussing. You are now moving, thinking on your feet, and holding the attention of the audience as you talk. The audience remembers things they see better than things they hear. Practice this type of response. When you are in a conference and are asked, "What wall section do you want?" Instead of telling them verbally, go to a board and draw the wall section. It will show that you know what you are talking about and will hold the audience's attention while they see exactly what it is you are drawing. There is no excuse for them not to understand your instructions. You are at the blackboard, in full view, holding up something in your hand and giving an exact answer, which is visible rather than just verbal. When a question is asked, determine if there is some detail, sketch, or list of items, which, if put on the board, would make the answer easier to understand and remember.

Presentation Run-Through

The presentation run-through must be a dress rehearsal, word for word, by each of the people taking part in the interview. You should time each part and actually go through each of the audiovisual presentations you plan to use. Keep rehearsing until your presentation is smooth and does what you want it to do in the allotted time. Now is the time to be critical of each other. Help each other to improve by smoothing out the rough spots and perfecting the delivery. Watch for annoying habits such as jingling coins, constant use of buzz words. Use your key-word notes in rehearsal, and it will be natural to refer to them in your actual presentations.

This series of rehearsals, prior to each interview, will build everyone's confidence and get rid of the need to depend on crutches or delaying techniques, which are used if you are nervous or trying to think of something to say.

Determine the points you wish to make and choose the exact words you wish to use to make them. People say repeatedly that they do not want to rehearse their interview in this way because the program will sound canned. Actually though, the rehearsal develops your answers to a point where you have a broad-scope response for any question asked and detailed statements to make. Stay flexible so that during the presentation you can pick your words to meet the situation, thereby tailoring the statement to the exact question and audience. The rehearsal will give consistency to your intent, even if you vary the words to suit the occasion.

Constant practicing prior to the interview will develop a cohesive team approach. Because you know exactly what you are going to say and exactly what the other team members are going to say, you can tailor your material to complement each other and assure a smooth transition from one speaker to the next, avoiding the old problem of each person saying the same thing in a different way. In a forty-five minute interview, you should only repeat information for special emphasis.

During the interview, you must be flexible enough to respond to questions from the audience at any time. This sometimes causes the speaker to miss a point he or she had intended to cover. But because you know what each of you had intended to say, you can learn to slip in a statement without interrupting the speaker to bring his or her attention back to the missed item. This technique of complementing each other and talking as a team is valuable, but learn to do it in a way that does not annoy the audience by its slickness, but rather says to them, "This is a strong team effort and all the members speak with knowledge of the project and as a team." This is a tough technique to master, but it is invaluable, once perfected.

A team presentation allows one person to listen to the audience and their questions at all times while another person answers the questions.

This person can watch the audience's expressions to establish if the message is getting through. He or she can jot down notes and then later on add missed items or clarify points by interjecting a statement or posing a question on his or her own.

Because we all have to brace ourselves for these presentations, we are sometimes impatient with the other speakers. We think they are talking too slowly or boring the audience and may show this in our own demeanor. Remember, the audience has not heard the material before and cannot absorb it as fast as you can. If things are going along as rehearsed within the prearranged time limits, everything is probably going well.

If all work at their material and delivery technique, the team will appear polished, and the presentation instead of sounding canned, will have a smooth, lively, professional, unbeatable quality to it.

Special Arrangements

Special arrangements must be made in advance to assure that all things you need for your presentation are in place before the interview. You might need a projection screen and a long extension cord. Avoid problems by bringing your fully equipped first aid kit, as discussed in Chapter 3. If you want to use a chalkboard, ask if you can meet in a room with one, or bring your own flip chart. Review the room size and number of people expected and the size and extent of the exhibits to the room. I went to one interview where the room was so small and crowded that the exhibits had to be set up in the hall. During the presentation, everyone was always in the wrong place and the exhibits missed their mark.

Decide what you will need, but cannot bring. Then in advance of the meeting, ask the owner to provide it. Check with him or her on the day of the interview to be certain everything is ready for you.

CHAPTER EIGHT

THE INTERVIEW AND BEYOND

This is the time you have all worked many long, hard hours to reach. The prospective job has probably been active for many weeks, and winning the contract may be vital to the health of your firm. But do not panic. If you have done your homework, your chances are good. Now is no time to worry about the things you may have omitted; rather, you must concentrate all your energies on doing the most professional job possible. After the interview, there will be time to analyze what you did and to note the good and bad points so that you can improve your subsequent performances.

THE PRESENTATION

Your Arrival

Allow plenty of time to get to the interview. Arriving an hour before the appointed time is not too early. You must have some time in reserve to allow for possible misadventures along the way. Give yourself enough time to make certain that, even if all the possible problems and mishaps actually occur, you still arrive before the appointed hour, relaxed and fresh.

If it is possible to get into the building early, find the room where the presentations will take place. Observe the seating arrangement. How many people will be there? Who are they? This will allow you to make up your seating chart in advance. Try to observe a speaker and the audience's response to him or her. Are they asleep; are they just sitting there; are they wide awake and enthusiastic? It is always good to know something about your competition. If you arrive early enough you will have time to make some last minute adjustments in your presentation in reaction to what you have observed.

As you look into the room, try to catch the eye of your official contact. This person then knows, not only that you are there, but that you are early and obviously excited about the interview. This will put pressure on him or her to try to close the current interview as soon as possible. The quicker the preceding architect, or engineer is excused, the less time he or she has to influence the board. You might get an opportunity to start your presentation early, but it is not a good idea to appear pushy in this respect.

Once you have made your presence known, you should stay within hearing range, but out of sight. Listen for the audience's reaction to your competition to give you a lead as to what is happening, but do not stay long enough to become distracted. It is more important to use this time to prepare yourself mentally for your own interview.

Getting Ready to Enter the Room

Organize your entrance into the room to convey the impression of professionalism and self-confidence. Also organize your team so that each member is responsible for certain items, which he or she should carry into the meeting room and put where they are to be used.

Everything, including the slide projector, should be unpacked outside the meeting room and set up to make sure that everything works. When everything is unpacked, pile all your boxes, cases, and coats neatly in the corridor. The senior member of the team should enter the room first and begin greeting the people. Out of the corner of his or her eye, this person should check to make certain everything is in place. Nothing is more embarrassing when one moves to begin the slide show than to find that the projector is still out in the hall. These embarrassments distract your audience's attention, which once lost is hard to win back.

Your interview really begins when you enter the room. Be organized, yet relaxed. Move smoothly, efficiently, and confidently, for everyone is watching you. The first thirty seconds of observation of your firm in action may well be the impression they remember most clearly.

Avoid meeting the previous firm. If you do cross paths, be polite; greet them quickly, then continue the work at hand. Prolonged contact with the previously interviewed firms will distract you and consequently give them an edge. You want to create a different impression, but too much socializing with them in front of your prospective clients give them the impression that you are all alike. Be different. Be in command of yourself at all times and avoid being neutralized by the other firm.

As you greet your prospective clients, note their attitudes. Are they sleepy and tired after a long dull session? Are they keyed up and excited? Their attitude is the indicator for your first actions. If they are sleepy, open a window and let in some fresh air. Ask them to help you set up your equipment. Wake them up! Get them on their feet and involved in your preparation. If you do not succeed in waking them up you will never arouse them during the presentation. If they are asleep, do not start your presentation with slides, or you will put them into an even deeper torpor!

Unless you have a specific reason and prior approval, do not rearrange their chairs. Most boards are very strict followers of the pecking order, and when a person is elected to membership, he or she is elected to fill a certain chair and that chair is placed in a certain position at the table.

As soon as you enter the room, everyone should introduce himself or herself and get a conversation started. If you have someone with you who has never been involved in a group interview, instruct him or her carefully in this procedure. Greet the board members individually and warmly. Tell them your name, the name of the firm and your position. At this point the conversation should be general and not related to interview material. You must convey the feeling of self-confidence and that you are excited to be here. You must treat the person you are greeting as if he or she is the most important person in the world.

Your setting up process should have been so well rehearsed that it appears effortless. You can carry on a conversation as you set up if you know what you are doing.

As your team is completing the setting-up process, your leader should take the leader of the board aside and discuss the format of the meeting and the introductions. This is a good time to show the leader of the board your seating chart and ask him or her to help you fill in the names of his or her people. The board leader will be more than willing to do this and once again you are exhibiting your concern for organization and details as well as making the interview personal enough to allow you to call each person by name as you answer questions. Your seating chart will be of vital importance to you.

Seating Arrangements

The seating arrangements will tell you a lot about the members of your audience and the style of presentation to which they will best respond.

1. FORMAL ARRANGEMENTS. Here chairs are set up in a prearranged order. This order is determined by the authority or tenure of the person occupying it. In this arrangement, the people sit erect with their standard office-prepared notebooks and manuals and lettered name plates in front of them. They are organization conscious and will respond well to a firm that is highly organized.

   They will naturally tend to respond more easily to an older, more mature person than they will to someone younger. They want you to present the facts in broad-scope terms, and their questions will get into the facts in-depth. They insist on exact answers and will hold you to them once they are made.

   You may have the feeling they are trying to discredit you when they start their questioning, but this is their way of getting to the facts.

   Once you have won their confidence and they believe your answers, your progress with them will be good. Do not hold back information or mislead them. They will discover it and call your attention to it.

2. CLASSROOM SETTINGS. These are, as the term indicates, chairs arranged in rows, with the teacher's desk in front of the room. This is the setting in which introductions and presentations are made. The chairs generally have writing arms, which seem to suggest that the board members are saying, "We want to learn in order to make the right decision." On each desk arm will be a pad of paper, pencil, and evidence of note taking from the previous interview. This audience will have done its homework and will ask specific questions of each architect and

then compare the answers as they review all the presentations. They will appear eager for knowledge and will take notes constantly.

The best way to communicate with this group is to assume the role of a professional teacher and present to them meaningful information of a specific nature. Do not use meaningless jargon. They want specific facts about your plans for their project. This is an excellent opportunity to use the chalkboard or flip chart. Classroom settings are excellent and can lead to good question and answer sessions.

3. THE LECTURE ROOM FORMAT. Generally, this arrangement is used when there is a large group of people for the presentation. The audience sit in chairs without a writing surface in defined, formal rows. The presentation takes place from a raised platform, generally at a speaker's podium or table. This is a difficult situation to work in because you are looking down on the audience, making it difficult to maintain constant audience contact. The size of the audience and distance from them often require the use of a public address system, resulting in a loss of personal contact.

   Since it is awkward for the audience to take notes, the questions become very strained, because the person asking them feels he or she is on display. Avoid the lecture presentation if at all possible. If you cannot avoid it, recall the problems and find a way to overcome them. A simple suggestion, and one that has worked for me, is to take the microphone and move away from the podium. Sit on the edge of the stage and talk to the audience on their level.

4. DISCUSSION GROUPS. This is a casual, seemingly unstructured meeting mind. The governor of Vermont once invited members of my firm to meet in his house along with several key people from the State Education Department to discuss new ideas that might reduce the cost of school construction. By meeting in his house, we all were more relaxed and the conversation was freer and more productive. But do not be deceived into ignoring the structured chain of command, which still exists and which you must recognize at all times. Do not let the soft chairs and the informal talk fool you into becoming too personal. You are still a visitor and are being judged as a professional outsider and not as a member of the group. It is unwise to accept any alcoholic

beverages while involved in a meeting of this type or, in fact, at any meeting. A discussion group makes it easy to talk. They want specific information. Your prepared presentation will have exact details for answers, not jargon. Keep track of the questions, and if at the end some material still has not been covered, use your outline to cover it then. Be casual, but do not lose your professional poise or position.

How to Hold the Audience's Attention and Interest

Do not sit behind a table. If you do, you will blend in with the furniture and the audience's eyes will wander and focus on other things. Standing behind a table commands their respect and will hold their attention longer. When you move in front of the table, you will not only hold their interest because you are standing, but you will attract additional interest because you have shortened the distance between you and the audience. Sometimes it is permissible to sit on the front edge of the table or desk. This helps to develop an informal atmosphere, but do not let this be interpreted as a sign of your boredom. Never sit on the table or desk at a formal interview.

Standing up commands attention, but your standing up and writing on a chalkboard or flip chart command and control the audience's attention even longer.

Wear stylish, well-cut clothes. We were invited to an interview for a large vocational education center in the Boston area. The board interviewed about twenty-five firms over a two-month period. This seemed a sizable task for the owner. How could he remember the firms, let alone the people who represented them? A week after the interview, my partner asked one of the board members how he kept the firms straight. He answered, "At your interview, you wore a plaid suit and a brown tie." He showed us his notes. He had described unique items from each person's wardrobe.

Another idea for commanding attention is to control the size of the meeting place. If it is large, set up some visual barriers behind the speaker and down the sides of the room. By angling the barriers, you can direct their eyes to the speaker. The lights behind the barriers should be dimmed or turned off. Reduce the overall room lighting for slide presentations, but never turn the lights completely off. Leaving one or two lights on helps your audience stay alert. It also encourages questions during the slide show, which they would not ask, if the room were dark.

Your presentation should appear informal. Rehearsals will help you to achieve this. Keep a professional attitude at all times, but do not keep reminding the audience of your professionalism, it will show if it really exists.

In all your conversation, use only positive statements. Avoid negative references. When asked a question that would ordinarily be answered in negative terms, try to give positive aspects, too. For example, when asked, "Is the cost of construction going up because of inflation?" An affirmative answer indicates added cost to the project and additional expense for the owner who begins to think, "Is it really worth it?" By using a little ingenuity, you can answer, "Yes, it is, but let me tell you what our firm has done to counteract inflation, the result being that the average cost of construction of our firm's projects has gone down."

You have won that owner's respect for meeting some of today's economic problems head-on. He or she now knows you will shepherd his or her money, too.

Do not bore your audience with a proliferation of unnecessary detail. When someone asks you, "How are you?," and you go into all your aches and pains, they regret the question. Not only will they try to prevent you from giving the details, but worse, they will never ask you another question. When asked a question at an interview, answer it simply and directly. If they want more details, they will ask more questions.

It is important to get the audience to ask questions. Questions break down the barriers between the speaker and the audience. Questions relate the material being presented to the audience's own project. If your audience does not ask questions, you have failed to excite them or appeared aloof from their concerns. This is a sign that you should relax your presentation style.

At the beginning of a presentation, a speaker always says: "You may interrupt me at any point if you have questions." But having said this, they launch into their speech, without pausing long enough to give anyone a chance to ask questions and without recognizing a look on someone's face indicating the desire to ask a question. This kind of speaker would really prefer not to have questions, fearing that since they are not part of the prepared material, they may throw him or her off balance. But if you slow down and give your audience a chance, you can use their questions as a guide to direct your material toward their specific areas of concern.

What do you do if they don't ask questions? This can be a real problem. Suppose you have done everything recommended to get the audience excited enough to ask questions, but the questions still do not come? Have a list of questions other audiences have asked at other interviews. When you reach a dead spot, have one of your team members ask one of these questions. It can be prefaced by saying, "At an interview last week, the audience asked this question about the relationship of something we said to their project. Perhaps it is valid here tonight." Watch the audience at this point. If one or two heads in the audience nod in agreement, answer the question, talking directly to the nodders.

Once you get them to ask a question, answer it. By this, I mean listen to the whole question. Take a couple of seconds to think about what was said and determine just exactly what they are asking. We sometimes have preconceptions about the question that prevent our answers from reaching their target. This is another reason to have a "listener" at the interview. One person can listen to the question and make sure the speaker answers it. Watch the face of the questioner. If it remains expressionless once you answer, it means you have missed the point and should try again. If the person starts nodding and smiling, it indicates a satisfactory answer. If your listening partner feels more information is needed in your answer, it is valid for him or her to add further comments after you have finished. This will encourage more questions.

Your preinterview research told you the level of construction knowledge to expect. Pitch your presentation at this level of sophistication. If you use the audience's vocabulary, it must sound as if it were the level of conversation you use in your everyday life. Never let them think you are coming down to their level. When you use a word that is not familiar to them, take time to explain it so they will always know just what you are saying. Never lecture the audience, but talk to, not at, them.

If you become sidetracked by a question, check your outline to bring you back on track. Never let questions divert you from your presentation, which must end on a high note. You want to leave your audience excited when you finish.

A word of caution: the more presentations you make, the more apt you are to become overbearing. As you near the room for the presentation, your adrenalin starts to flow, not in fear, but in anticipation. This can be good if it is controlled; uncontrolled, this can cause trouble. You may begin to think of yourself as an inspiring teacher-lecturer. You may then develop a tendency to control the presentation, without giving the other team members a chance. You will stop listening and responding to the audience. Watch yourself closely to avoid this trap and watch each of the others on the team to head off any member you see moving in this direction. If it is allowed to go on unchecked, your effectiveness will drop drastically.

Watch for Signs

Watch for signs from the audience as you give your presentation. This should be done as a team. Those not talking should be listening and watching the audience.

Learn to read the faces of people. Ask yourself what their expressions mean. Is the speaker not answering their questions? Are the answers too long? Is the audience becoming bored? Is it time for a new speaker? Learn to read and understand body language. Several good books are on the market about this subject. Watch the body signs. They give you vital information. Understanding body language will help to make you a more effective speaker and a more efficient and successful salesperson.

Use your team to perfect your presentation. While one person is talking, the other is listening and looking. Learn to interrupt each other effectively and to operate as a team. Watch each other and be sure the material is fully covered and the audience's signals are read and responded to. Learn to take criticism from your teammates since this is vital to your growth as a speaker. Be objective in giving and receiving criticism.

The Interview

You have forty-five minutes to present your firm and to stand out sufficiently from the competition to get the job. Forty-five minutes is not very long; so make every minute count.

Keep all your handouts, except the outline, until the end. If you pass them out first, the audience will be leafing through them when you want their attention. As you start, give them the one-page outline so they can see what you are covering.

Next, introduce yourself and the other people with you. This would be a good time to use your flip chart. Write plainly the names and positions of each person and what their relationship to this particular project will be.

Your names and positions should also be listed at the top of the outline. Hold your business cards until the end of the interview and give them out with the other material. Handing out cards when you enter the room and while you are making the introductions leads to confusion.

As you go through the presentation, make the transition from one subject to the next and from one speaker to the next as smoothly as possible. The ultimate aim is for the whole presentation to appear as if one person made it.

When the main presentation has been completed, the leader should take over and provide a summary of material covered. This is another opportunity to use your action aids as you put a listing of all the major points covered in your presentation on the blackboard or a flip chart.

After the summary, distribute the handouts. This is a delaying technique to give them time to think and get ready to ask questions. It also gives you something to do while they are thinking. The hoped-for results will be more questions, allowing you to reopen the conversation and continue the interview on a personal level.

Your handouts should include your company brochures with business cards attached to the inside covers as well as the additional material previously described.

Now that your audience has had a chance to think and regroup, it is time to go into your closing. Unless this is the final interview, your goal is to get them committed to another interview or other contact.

A good next contact is a visit to one of your projects or to your office. If the interview went well, but you feel they would like to talk more, suggest another meeting with them. They might take you up on it.

Suggest they contact your references. Offer to meet with them to develop further the scope of the work or to discuss the preparation of a contract proposal. But be careful not to force any of these ideas. Offer them, but if they show no interest, move on to the next thought. Often an idea planted at the interview is acted on the next day.

Your Departure

When it is obvious the interview is over, gather up your material quickly and efficiently, and let the audience help if they offer to. Move everything out into the corridor before you start to disassemble and pack it. This will clear the room quickly and remove the confusion of packing from their view. It also frees the room so that they may discuss your interview privately. Depart smoothly and efficiently. Look as organized, professional, and confident as when you came in.

FOLLOW-UP

Next Day

Wait until about ten o'clock the next day for their phone call. In this business, no news is bad news. If they have selected someone else, that good friend on the board who used to call you will be hesitant to call. They will let the leader of the board send you a "Dear John" letter. It is up to you to keep his or her friendship. Give the person a call, saying you are interested in how your presentation went last night and in the status of the selection process. If there is a long pause, prepare yourself for the bad news.

If the person tells you someone else was selected, be warm and professional. Admit you are disappointed, but let the person know that it is neither the end of the world nor of your friendship. Be the normal enthusiastic person you were when you started developing the project. Above all, do not lose this contact. This client will probably build other buildings in the future and you want to be invited to subsequent interviews.

One of the first prospective projects for which I represented my firm was a large high school. I put everything into the development of the prospect and felt that the job was mine. My contact in the superintendent's office confirmed my feelings. On the night of the selection, the competition outmaneuvered me and my firm ended in second place. I was upset because my hopes were high and the building was already designed in my mind. I called my contact and got the bad news. I was disappointed and so was he. I regrouped, as my partners had taught me to do, and started prospecting the next project I thought this prospect might have. I stayed in touch with my contact, with the result that my firm was brought in as advisers on the school's next project and was subsequently hired to build a flexibly planned school for them.

When you call, if you are told that neither your firm nor one of the others was hired, find out from your contact how your presentation went and if you covered all the material he or she felt you should have. If not, find out what you neglected. What is the status of the project now? Is the owner going to talk to additional firms? Are visits being planned to some buildings? Will several firms be chosen for another interview? Are you one of the firms?

Once you have found out what is happening, see if your contact will give you any suggestions about what you should do to get your firm selected. Encourage him or her to set up a visit or another meeting. Each day that passes after your interview is crucial. The image you made with the board starts to fade. The competition has a chance to counter your presentations.

Sit down with your prospective teammate and review in detail the past interview and your phone call. Be realistic in your evaluation of the interview. Try to decide what happened. What did the competition do that is affecting us? What can we do to improve our position? Where did we miss? What are our weak and strong points?

After you have fully analyzed the situation, put on paper what has to be done. This is an essential part of the follow-up process. The real key to your success is your proper diagnosis of the situation and the development of the right prognosis. Use your imagination in deciding what to do. The real situation may be quite different from the obvious one.

You must then prepare a letter to be sent to all those who were present at the interview. You should thank them for the interview and touch on some of the key points you want to emphasize. Include any material your analysis showed should be brought to their attention. Close with a statement of your willingness to meet with them again. Also invite them to visit some of your projects and your office.

The morning after an interview is anticlimactic, and it is a temptation to involve yourself in other projects. But do not give up on this prospect now. You have spent a great deal of time and money to get to this stage, and you are now in the home stretch. Remember, no news is bad news, and the longer you stay out of contact, the more likely they are to forget you.

A Week to Ten Days

This is an extremely critical time in the closing of the contract. One of the first things you should do is to revisit all the key people present at the meeting to see if they have any unanswered questions. You can hand deliver any additional material you feel important to the situation. Get them to talk about their project, not necessarily about your firm. During the conversation, they will reveal what they thought of you and your firm and the kind of presentation you made. If they avoid the subject, it is a bad sign that indicates your chances are fading. If they think highly of you, they will volunteer information about your presentation, about the other firms, and the status of the project.

This is the time to get them to start building their project in their imagination. You have to help them feel they are building something personal. A good initiator of this process is a schedule, which can be used to show all steps necessary for them to occupy their building on the date they have chosen. This will emphasize that they should start as soon as possible to meet the deadline.

Obtain another meeting, if at all possible. This could take the form of a discussion of a special subject directly relevant to the project, such as zoning or new regulations for the handicapped or a presentation to them of additional information on your firm you were unable to cover at the time of the interview. The meeting must have a valid purpose, not be an excuse to prolong the process.

Keep It Going

This is an easy time to become discouraged, especially if your position seems to be slipping or the decision is being postponed for several weeks or months. Keep analyzing the prospect. Ask yourself, "What are they really saying? What did we do? What should we have done?" Count the votes and decide your position in the voting. Decide what has to be done. Then do it. Above all, do not indulge in self-pity or depression because you think someone else will get the job, which is possible. Get out of the office; revisit the owner. Think positively.

Keep going, but be cautious of wearing out your welcome. The competition is probably doing the same thing. Some people involved may become irritated at the process; however, some board members may enjoy the attention. If you stop to visit someone and that person appears irritated, do not take it personally. You may be the third architect to visit him or her in three days! Excuse yourself politely with an apology for bothering him or her. Do not necessarily stop all your visits because of the results of a bad one. Visit a retired member of the board, who will almost certainly welcome you.

On a prospective high school job in southern Vermont, I became involved in other work after the interview and before I realized it, two weeks had gone by. My colleague brought this to my attention and told me to drop everything and make contact with the board chairman. His first statement to me when I called on him was, "I'm glad you finally came around. I really would like your firm to build our school, but we were beginning to feel you weren't interested. During the past two weeks, the other firms have visited all of the board members at least twice!" I was lucky this man was frank with me. We got the job.

Do not stop trying until you, or someone else, has the contract. Even then, keep this client on your list for future prospects.

The Closing

Although days, weeks, or even years may have gone into the preparation for the closing, many people do not know how to close the deal. They may not recognize the signals that the client is ready to sign; or they may fail to understand the psychology of the process, what the client seems to be saying, and what he or she really is saying. It is therefore important to examine the appropriate techniques for setting up the closing, as well as to understand signs typically raised by the client that indicate he or she is ready to close. This will allow you to understand the process and achieve a successful closing, instead of letting the moment pass without realizing that the client was ready to sign.

Timing, the most important technique to master, is one that cannot be taught, but only learned through close observation of many closing situations as they occur. If you try to close too early, you will be considered pushy and marked as a high-pressure salesperson. If you try to close too late, someone else will have made the sale. The right move at the right time is essential to the contract.

To master the art of timing you must learn to feel the vibrations and energy that emanate from the people with whom you have contact. I cannot really describe this feeling, but once you experienced it, you will recognize it when it happens again. You must be prepared for the possibility of a closing each time you meet a potential client, and each meeting should take you one step closer to that point. A closing seldom happens on the first encounter, but it has been known to happen very quickly.

If you repeatedly get to the point of closing the deal, but never emerge with a contract, you should admit that you cannot handle the situation alone and bring in help from your office.

The following is a series of standard techniques which can be used to assist a closing.

1. STRAIGHTFORWARD APPROACH. You can sometimes simply ask the question in a polite and low-key way: "Do you want to hire us?" The owner can only say yes or no. If he says yes, you have the job. If he says no, you then can ask why, and he will give you some reasons. This gives you a chance to reopen the conversation and to continue your presentation.

2. REALIZE YOU HAVE THE SALE MADE. If the client is sold, recognize it and stop talking before you talk yourself out of the sale.

3. THE SILENT TREATMENT. This is a technique that should be used only as a last resort. When you have finished your presentation, stop talking and just stand there. Within about twenty-five to thirty seconds, the silence will become awkward, and someone will say something. The

one who starts talking first usually loses. For example, if the owner is the one to break the silence, he or she will usually either commit himself or herself to you in some way or give you a reason why he or she cannot. This allows you to reopen the negotiation. If you are the first to talk, this usually signals your exit. Try the technique first on one of your friends and see what develops.

4. ALTERNATE CHOICES. You can often lead a potential owner toward a contract by designing the building for him or her verbally during your conversation. You can ask him or her the following questions.

   a. What do you like for windows—casement, awning, sliding, aluminum, steel, wood?
   b. What kind of cabinets do you like—wood, metal, plastic, contemporary?
   c. What kind of carpeting do you like and where do you like it installed?

   By going through a complete list of all the basic items in the building you will acquire sufficient understanding of the program, area, and budget to say: "Well, it seems to me that this building should cost approximately 'X' dollars." The verbal design of a theoretical building is often successful in closing the deal.

5. THE DOOR-KNOB FEINT. As the client is ready to leave, or the interview is definitely over and lost, it is sometimes possible to say: "Well, I guess I have not convinced you to hire us to design your buildings. Would you mind telling me why?" The potential client will often be flattered by this approach and will start to tell you what the problems are. This in turn will allow you to reopen the conversation.

6. THE BEN FRANKLIN BALANCE SHEET. The owner will often say that he or she is not ready to buy or to build. Your response can then be: "All right, let's try a little exercise. Let's see what we need to do to get your project underway." You develop a balance sheet with two columns. The first column is marked "plus," and the second "minus." You start off by listing all the pluses such as Do you have the land? Yes. Do you have a program? Yes. Then list all the other items to which you

are sure the client will respond affirmatively. After you have thirty or forty items in the plus column, hand the paper and pencil to the client and ask him or her to list all the items on the minus side. The person will usually come up with only two or three. The overpowering pluses will often outweigh the minuses, which you now also know and can attack one by one and move on to winning the contract.

7. THE PUPPY DOG APPROACH. This is similar to the little boy who brought a dog home, and having gotten the dog inside the house, then asked his mother if he could keep him. Who could refuse to keep a puppy as cute as that? In prospecting, a similar result is often achieved by bringing the owner some material tailored to his or her specific needs, with his or her name on it and the project spelled out in some detail, including scheduling, budgeting, sketches—whatever is ethically sound. Provide enough personal material to give the owner something to think and dream about. This will make the owner feel somewhat committed to you.

When an owner says, "I want to think it over," this usually means he or she will not call you back. "I want to think it over" should be countered with "If you are still thinking it over, it must be a very difficult and complex problem. Can I help you with it?" The owner will often say yes to this, but even if the response is no, he or she will often list the problems. And once the problem is out on the table, you can attack it.

Project managers often tighten up when they reach the closing, and the prospective client can often detect this tension and interpret it as pressure. When this happens, the success of the sale is jeopardized, and the closing is greatly endangered. Another common fault is inadequate preparation and faulty presentation. Neglect of this kind makes it difficult to close.

Success depends on the ability to recognize when the owner is ready to sign, and respond immediately with the proper technique. Prospective clients will often signal you that they are ready by changes in expression, by getting up and participating in the presentation or by their general enthusiasm. If the client group looks worried, perplexed, confused, or sits with arms and legs folded, you have a problem and need to change your technique to solve it. It is most important to get the clients involved in the presentation, to get them up, walking around the room and personally involved in what you are doing. The most important signals are the questions they ask. Listen carefully and have someone there to write the questions down so that you can analyze them properly. The more specific the questions become, especially when you get into the areas of fees, time schedules, and costs, the nearer they are getting to signing.

If the client is clearly not ready to sign a major contract with you, respond by identifying his or her immediate needs and offer the person a short contract. For example, if the client is not ready to hire you to design his or her building, perhaps you can get the person to give you a small contract for programming or site selection.

You should never exhaust all your selling ideas at the first interview, because if you are called back again, you will need something new to talk about. Moreover, if you present all your ideas the first day this will usually overpower the client and leave him or her greatly confused.

In your discussions with the prospective owners, always try to ask questions they can respond to affirmatively. In this way a receptive and positive attitude is generated that will help to develop a favorable climate for negotiation.

If you start by helping your prospective clients make the minor decisions that are easy for them, they may be led almost imperceptibly toward the major ones.

It is a major decision to hire an architect or engineer, and you must develop a whole series of questions and decisions to lead your client gently but firmly to this point.

It is often a good idea to bring in help for the closing because a new face, new expertise, or a new expression of familiar words may be all that is needed to clinch the agreement. If you have a second person at the closing, one person can listen to the exact questions while the other answers them, giving you an opportunity to correct any mistakes.

Do not be afraid to switch the course of the conversation if you see that the course that you are on seems to be putting your potential client to sleep. Be responsive and alert to the dangers of getting caught up in a standard format.

What do you do if the prospective client fails to sign at the proper moment? Remember that you will make many more sales presentations than you will sign contracts. Even when you get your "Dear John" letter, respect the decision and continue to keep in touch. Someday in the future the owner may want to build something else and at that time he or she may well come to you.

CHAPTER NINE

IF YOU LOSE A SALE

HOW TO RESPOND

You sit at your desk the morning after the biggest and most important interview you have ever had. You feel you did everything just right and the interview seemed completely positive. You arrived at your office this morning, certain that you had won the contract. You sit at your desk anxiously awaiting the time when you can call your contact on the board to find out the results. You get up your nerve and make the phone call. As the phone rings, you wait with baited breath. Then the answer. You say hello and you immediately notice a long pause as your contact recognizes your voice and tries to think of something to say. You have lost the contract and he or she does not know how best to tell you the bad news. Get yourself ready and make it as easy for the person as possible.

Accept the loss gracefully; try not to show too much emotion. It is best to make this phone call as short as possible.

IMMEDIATE RESPONSE

Do Not Be Ashamed of Failure

Do not be afraid of failure for there are more failures in this business than there are successes. If you are afraid of losing a sale, it will show in your attitude with prospective clients. You will push too hard. This does not instill confidence, but rather indicates panic. Look for the ratio of successes to failure in your past record. If you can consistently land two contracts out of every ten interviews, this is a good average.

An example of pushing too hard occurred early in my career of project management. Shortly after the firm had opened an office in Burlington, Vermont, and hired me to run it, I uncovered in my hometown a prospect of a new $5-million school. I felt almost entitled to this job, because I lived in the town and felt that I would do a better job than any of the other firms being considered. Also, since this was the first major prospect I had, it would be a substantial help in launching the firm in New England.

Everything went very smoothly at first. I had an interview with the prospect, and all indications seemed to point in the right direction. But then problems rose. Instead of taking these problems in my stride, analyzing them, and planning counterstrategies, I let the situation become a battle of survival for me. As I look back on the days that followed, I can see that my prospecting activity became a matter of begging for the job. It never quite came to that, but my panic was no doubt evident in my eyes and facial expressions as I talked to people and as I saw the project slowly slipping out of my grasp. But this emotional involvement and the lost job taught me a lesson early in my career. I learned to keep such activities strictly on a business level and not to become personally involved.

A critical time in your prospecting life is immediately after you discover you have lost the contract. Do not be depressed over the loss. When you make the phone call to your contact and are told the bad news, listen carefully to the reasons, but do not comment. Accept what your contact says. Nothing you say will reverse the decision. In fact, any comments other than those reflecting disappointment will be taken as those of a poor loser. If the person does not volunteer any reasons, do not press him or her for them because this places the person in a difficult situation. Wait a week or two until your emotions have calmed; then talk to the person privately and ask for the details. But do not look back. The contract is lost and should no longer occupy your mind. Start work at once on the next project or prospect. Start working twice as hard as you usually do, and make visible efforts to change your thoughts and channel them into new areas. If you dwell on the loss, it will affect your entire attitude, and your negativism will be communicated to your staff.

Accept the fact that you lost the prospect, and resist the temptation to blame the failure on illegal activities or politics behind the scenes. I have often been tempted to take that approach. But in all the interviews my firm lost, it was always because of something we did wrong or something we failed to do. Never did I find any evidence of anything underhanded that caused us to lose. It would often take a year or more to find out what the reason was, but it always came down to the fact that the owners did not relate to us and what we said or that they felt someone else could do a better job.

Once you are well on the way to acquiring the next contract, it is time to pause and study the past failure. This is best done about a week after the notification of the loss, giving you time to get involved in new work and to become unemotional about the loss. This is the time to review, factually and objectively, the entire history of the prospect. This is best done with your prospecting partner so that together you can review all the details of the prospect and try to pinpoint when and where it went wrong. How did the prospect become known to you? Was it one you found on a cold call, or was it one a client referred to you? If it was referred to you by a client, consider going back to this person to see what he or she knows about the situation and why you lost out. But be aware that this might put your client in an awkward position. Another question to ask is what service did the prospective owners say they wanted when you first talked to them? Did this change during the course of the prospect? Did your interview reflect their expressed needs and wants, or did it relate to what you thought they wanted? Were you selling full services, when all they wanted was a feasibility study? Did they misunderstand your statement of qualifications as a hard sell? On a step-by-step, day-by-day basis, review what happened. Review each visit you made, who you talked to, what you said. How did the competition handle the prospecting period? What were they doing, who did they see, what must they have said? What was their key approach? Did they relate more accurately to the real need than you did, or was their strategy simply better? Two weeks or so after the loss is the time to renew contact with your acquaintance on the board. This person should be able to tell you where you went wrong.

The answers to all these questions should give you a good picture of what happened. When you have pinpointed where things started to go wrong, try to define the exact time. Was it during the initial prospecting period? Did you visit the same person too often? Did you say the wrong thing at the wrong time? Perhaps you pushed too hard. Did you fail to get out and get to know the people involved in the project? When you entered the room for the interview, were all the faces familiar, or unfamiliar? If they were all new, you did not do your prospecting properly. Some people think that the biggest problem during the prospecting period is making too many visits and wearing out your welcome. I have never felt this to be the case. If you are visiting someone too often, you will know it by watching that person's face.

The next area to review is the interview. Did your presentation relate to the required services, or were you too general in your approach. I remember on one job I lost, I went back to talk to the friendly board member who had been on my side all along. The person told me a competing firm had a unique service. This firm assigned a project manager to manage the project from start to finish as his main responsibility. The board member told me they liked this extra service and that it was in fact the deciding factor in the award of the contract to the other firm. I was amazed. This service was the whole key to our firm's success, and was something I had stressed during the preliminary prospecting, but passed over lightly at the interview. The other firm made it their main presentation and, in so doing, related to the owners' chief concern.

You must accept that before the interview another firm may have the job clinched and that the interview process is only a formality. Try to spot this during your prospecting, but never hint at it to an owner. Plan your interview accordingly and still give it a try. You might be able to out-perform the competition at the interview and get the owner to change his or her mind. Also, taking the long view, this owner will probably build again, as will his or her friends.

Another factor that may have cost the job is the post-interview follow-up. Perhaps you showed the wrong building, or the building you showed may have had a problem that bothered them even though you thought it minor. If an owner cannot make up his or her mind between two firms, he or she will look for anything that will make one firm stand out.

Competition is good for you. It makes people who win contracts perform. It makes the ones who lose update and modernize to become competitive so they will survive. The answer to failure is not, for example, to get state governments to legislate against awards to out-of-state firms, but rather to work to provide a service that is competitive.

Now that the replay is complete, it is time to record the summary. Record everything that happened and the sequence. List your observations about what went wrong, when it went wrong, what you did about it at the time and what effect your corrective action had. Finally, list all your suggestions for corrective action to be used in the handling of future prospects.

Recovery

If you review all your losses, you will probably find a trend will start to develop. Some of the key observations that should be made are these: was the time lapse from identification of the prospect to the time of award so long that you either wore out your welcome or lagged in your efforts? Did you fail to identify the true leaders on the board and spend your time talking to the wrong people? Did the key people not know you?

When you have identified these trends, assess whether the problems listed are now occurring less frequently. Have you taken steps already to correct the deficiency, and if so, how successful has it been?

Take time to study and analyze your competition. Who is your competition? Are you up against the same firms each time or do you find the competition is different at each interview? Who is getting most of the work in the area you are prospecting? If it is usually the same firm or firms, you must analyze them and their operation carefully, as well as their sales techniques. Take time out of your schedule to visit one or two of their recently completed buildings. What are the details of construction? What are the maintenance factors? Do the mechanical systems work? Are the owners happy? When you look at these buildings, make certain you judge them fairly and objectively. Get an owner you are working for to talk with you about working with your competition on his or her previous jobs. Talk with the owner about the services they gave and find out his or her opinions of them. But never get trapped into running down the competition.

Check the productivity and effectiveness of the people in your firm who are prospecting, particularly in relation to the number of calls made per week and month. Selling comes down to one main thing: you must make calls. If you sit at home you will not find any leads or close any contracts.

The last point in the recovery process is to review your goals. Are you reaching the goals you have established, or are you constantly falling short? Reexamine your short-term goals. How many man-hours are you spending each week and month prospecting? How does it relate to the dollar volume of new work obtained? How can you increase the amount of prospecting time and contacts? What new goals should you establish? What effect will the new goals and demands have on other work the managers must produce? What will the dollar effect be on the time and energy invested?

Architects or engineers involved in business development must make at least two contacts per day in the search for new work, which means charging one to two hours of time each day to new business. In addition, they must spend one full day per week outside the office making calls and trying to find leads and working toward the closing of contracts. To survive, a firm must set a goal for each manager of at least two major interviews per month.

Look at the long-term effect of your established goals. Will you and your other managers require back-up personnel to take on the routine tasks and thereby free them to have additional time to prospect? Should you consider the reassignment of some of your managers? If one of your prospectors is more efficient than the rest, consider making him or her available to other managers to help them uncover work in their territories or even reassign the person to a new area where business looks as though it will be growing. Use your key people to obtain new work as well as to train your other managers. Consider the need to reorganize the territory of each of the managers as well as the personnel serving the territory. Recognize that markets and needs are constantly changing and you must be attuned to today's needs and services.

As part of establishing a corrective, positive action program, make a serious attempt to do a market analysis of your area. What kinds of service have the owners required in the past year? What changes in service will they require in the next few years? Take a look at your service and determine if your present service, as it is now performed, will meet the needs of today's owners. Should you develop additional services and options to make you competitive and to give the owners what they want?

The key to recovery is to constantly analyze your prospecting wins and losses and take enlightened action to capitalize on the strengths and correct the weaknesses.

In the whole business of selling architectural or engineering services, especially after losing an important contact—one can never overestimate the power of positive thinking. Positive thoughts generate positive action. This takes a conscious effort, since the negative approach is much easier.

But if you allow the negativism to persist, you can expect continuing failures. At one time I worked with an architect who refused to view the future with excitement or to work positively and energetically toward success. He seemed intent on dwelling on past failures. Whenever we planned an interview, he enumerated all the things that could go wrong and finally depressed us all so much that we felt like canceling the interview. Some of the things he forecast did in fact happen, but this was because he had talked himself out of being successful in the interview before he got there.

So once you have lost an interview, put it behind you. Concentrate on something new. As soon as you become absorbed in a new prospect, it will generate its own momentum, and you will be working toward the next interview. But a realistic appraisal of what went wrong the last time will improve your chances of success in the future.

Remember that in the process of prospecting there will be many long hours and disappointments. There will be more losses than wins. In other words, it will be easy to fall prey to self-pity and defeatism. But if you remember only one thing from this book, remember the power of the mind. Thinking makes it so. A positive approach and an aggressive and sincere attack on any problem or process and your attitude will quickly become contagious, and others will follow and help you in your cause, thereby assuring you success in your undertaking. Good luck and happy prospecting.

INDEX